D0202879

ERAU - PRESCOTT LIBRARY

THE ETTO PRINCIPLE: EFFICIENCY-THOROUGHNESS TRADE-OFF

The ETTO Principle: Efficiency-Thoroughness Trade-Off

Why Things That Go Right Sometimes Go Wrong

ERIK HOLLNAGEL
MINES ParisTech, France

ASHGATE

© Erik Hollnagel 2009

All rights reserved. No part of this publication may be reproduced, stored in a retrieval system or transmitted in any form or by any means, electronic, mechanical, photocopying, recording or otherwise without the prior permission of the publisher.

Erik Hollnagel has asserted his moral right under the Copyright, Designs and Patents Act, 1988, to be identified as the author of this work.

Published by
Ashgate Publishing Limited
Wey Court East
Union Road
Farnham
Surrey, GU9 7PT
England

Ashgate Publishing Company
Suite 420
101 Cherry Street
Burlington
VT 05401-4405
USA

www.ashgate.com

British Library Cataloguing in Publication Data
Hollnagel, Erik, 1941-
The ETTO Principle-- efficiency-thoroughness trade-off:
why things that go right sometimes go wrong.
1. Industrial safety--Psychological aspects. 2. Human engineering.
I. Title
620.8'2-dc22

Library of Congress Cataloging-in-Publication Data
Hollnagel, Erik, 1941-
The ETTO principle : efficiency-thoroughness trade-off : why things that go right sometimes go wrong / by Erik Hollnagel.
p. cm.
Includes index.
ISBN 978-0-7546-7678-2 (pbk) -- ISBN 978-0-7546-7677-5 (hardcover) 1. Performance technology. 2. Performance--Psychological aspects. 3. Industrial accidents. I. Title.

HF5549.5.P37.H65 2009
658.3'14--dc22

2009005301

ISBN: 978-0-7546-7677-5 (hbk)
ISBN: 978-0-7546-7678-2 (pbk)

Mixed Sources
Product group from well-managed forests and other controlled sources
www.fsc.org Cert no. SA-COC-1565
© 1996 Forest Stewardship Council

Printed and bound in Great Britain by
MPG Books Ltd, Bodmin, Cornwall.

Table of Contents

List of Figures

List of Tables

Prologue

What is arguably one of the most influential papers in contemporary psychology starts rather tantalisingly as follows:

> My problem is that I have been persecuted by an integer. For seven years this number has followed me around, has intruded in my most private data, and has assaulted me from the pages of our most public journals. ... The persistence with which this number plagues me is far more than a random accident. ... Either there really is something unusual about the number or else I am suffering from delusions of persecution.

(The paper in question is George Miller's 'The Magical Number Seven, Plus or Minus Two: Some Limits on Our Capacity for Processing Information' published in 1956. This paper introduced to the general public the notion of limitations in human short-term memory and attention, and proposed as quantification an integer that since then has become legendary – but also widely disputed.)

My problem is not that I am persecuted by an integer, but rather that a certain idea has stuck in my mind, leading me to see examples of it everywhere. That in itself is not so strange. We all know that the moment we start to think of something – or buy something such as a new car or a new gadget – then we also begin to notice instances of it everywhere. This is what psychologists call the phenomenon of selective attention, i.e., that the way we look at the world is heavily influenced, or determined, by our expectations and preconceived ideas. The phenomenon is aptly captured by the adage that 'if your only tool is a hammer, then everything looks like a nail' (attributed to the American psychologist Abraham Maslow, but also said to be a Japanese proverb). In the world of accident investigation it has been expressed as the *What-You-Look-For-Is-What-You-Find* or *WYLFIWYF* principle, to be described in Chapter 5. In other words, our (current) understanding of the world heavily influences what we pay attention to or notice, both the larger 'picture' and the nuances. In consequence of that it also excludes from our awareness that which we are not prepared for, that which we do not expect, and that which we are otherwise unwilling to 'see.'

My problem is that since I started to think about the efficiency-thoroughness trade-off (ETTO) principle as a way to make sense out of what people do, I seem to find examples of ETTO everywhere. This is not something that I do intentionally, but the efficiency-thoroughness trade-off principle is seemingly ubiquitous. Indeed, the obviousness of the phenomenon is so strong that reason seems to demand that it should be questioned. But try as I might to eradicate it, it still persists. Writing this book can therefore, in a sense, be seen as a way to get rid of the ETTO demon, or at least to pass it on to someone else, like the *Monkey's Paw*. If that does not succeed, then I stand corrected and there is no such thing as the ETTO principle. But if it succeeds, then it may have significant consequences for how we perceive, analyse and understand human and organisational performance in general, and how we view the role that humans play for safety in particular.

Yet this book does not really describe anything new, in the intellectual sense that no one has ever thought of it before. As the examples, large and small, throughout the book will show, people – practitioners and experts alike – have for many years been thinking along the same lines and have expressed it in ways that are not too dissimilar from what is done here. The present text consequently does not and cannot pretend to be an intellectual breakthrough or even an innovation. It is rather a way of drawing together the experience from many different fields and summarise a wide collection of findings coherently, with a view to their practical consequences and their practical applications.

Everything around us changes and it often changes so rapidly that we cannot comfortably cope with it. In consequence of that, the descriptions that we make and use are never complete. This means that neither the situations we are in, nor future situations, can be completely described. There is therefore always some uncertainty, and because of the uncertainty there is also risk. If something is going to happen with certainty – or not going to happen with certainty, which amounts to the same thing – then there is no risk. (While philosophers may argue about whether something ever can be absolutely certain or known with absolute certainty, the rest of us can normally distinguish between what is certain and what is uncertain on a purely practical basis.) But wherever there is risk, there is also a need to understand the risk.

Although people and societies have tried to protect themselves against hazards and risks at least as far back as *The Code of Hammurabi*, the consequences of the rampant technological developments that we have seen since the middle of the 20th century – according to the Western way of counting, of course – have made this more necessary than ever.

Regardless of whether risk is defined from an engineering, a financial, a statistical, an information theoretical, a business, or a psychological perspective, the concept of uncertainty is a necessary part. This has of course been so at all times, but it is a quality or a characteristic that has become more important as the systems we depend on and the societies we live in have become more complex. Two hundred years ago, to take an arbitrary number, countries only needed to care about their nearest neighbours or their coalition partners, and were largely independent of the rest of the world. The same was the situation for institutions, companies, societies, and individuals at the appropriate scale. Neither financial nor industrial markets were tightly coupled and events developed at a far slower pace. Today the situation is radically different, basically because we have been caught in a self-reinforcing cycle of technology-driven development. In 1984, the sociologist Charles Perrow, who will be mentioned several times in these pages, observed that 'on the whole, we have complex systems because we don't know how to produce the output through linear systems.' The situation has not become any simpler in the years since then. Because there nearly always is too little time and too much information relative to what needs to be done, it is inevitable that what we do will be a compromise between what we *must* do in order not to be left behind, and what we *should* do in order to avoid unnecessary risks. In other words, a compromise or trade-off between efficiency and thoroughness.

About the Style of this Book

This book is not written in the style of an academic text, and therefore dispenses with the references that are normally found in such works. This has been deliberately done to make it an easier read and to make it more accessible to people who, for one reason or another, are reluctant to start on a conventional textbook or work of science. However, in order to meet some modicum of academic credibility, each chapter will conclude by a short section that provides links to the most important references and literature. This section can safely be skipped by readers who have no interest in such matters. On the other hand, the section may not provide the full set of scientific references for the more inquisitive reader, and therefore in itself represents an ETTO.

Sources for Prologue

When George Miller published the paper about 'The Magical Number Seven, Plus or Minus Two' in 1956, he offered a powerful simplification that quickly spread beyond experimental psychology. The 'magical number' was instrumental in disseminating the idea that the human mind could be described as an information–processing system, with all that this has led to. In relation to safety, the idea was later used as support for the idea that the human was 'just' a complex machine, and that this machine could fail or malfunction in the same way that other machines do.

According to the information–processing viewpoint that George Miller helped introduce, humans receive information from the environment and then process it. Humans are therefore described as passive or reactive 'machines.' Philosophers and psychologists have, however, long known that such is not the case. Humans actively seek information rather than passively receive it. In psychology and cognitive engineering this is represented by a perception-action cycle. This describes how our current understanding and expectations, sometimes called 'schemata,' determine what we look for and how we interpret it. In other words *What-You-Look-For-Is-What-You-Find*, whether it is on the level of individual perception or a collective activity such as accident investigation.

The Monkey's Paw is one of the classical horror stories. The modern version of it, written by W. W. Jacobs, was published in England in 1902. The basic plot is that three wishes are granted to anyone who possesses the paw of a dead monkey, but that the wishes come with a terrible price.

The Code of Hammurabi was enacted by the sixth king of Babylon, Hammurabi, about 1760 BC. One part of it describes what is known as *bottomry* contracts, a type of insurance for merchant ships. It essentially means borrowing money on the bottom, or keel, of a ship. The money would be used to finance a voyage, but the repayment would be contingent on the ship successfully completing the voyage.

In 1984, the US sociologist Charles Perrow published a book called *Normal Accidents* which argued that accidents should be explained as the near inevitable result of increasingly complex and incomprehensible socio-technical systems. This soon became known as the *Normal Accident Theory* (*NAT*), accepted by some and disputed by others. Further details will be provided in Chapter 1.

For the record, the first public airing of the ETTO principle was during a panel discussion at the 8th IFAC/IFIP/IFORS/IEA Symposium on Analysis, Design, and Evaluation of human–Machine Systems that took place in Kassel, Germany, 18–20 September 2001. The first printed reference is E. Hollnagel, (2002), *Understanding Accidents – From Root Causes to Performance Variability*, in J. J. Persensky, B. Hallbert and H. Blackman (eds), Proceedings of IEEE 7th Conference on Human Factors and Power Plants: New Century, New Trends, 15–19 September, Scottsdale, AZ. A description can also be found in Chapter 5 of Hollnagel, E. (2004), *Barriers and Accident Prevention* (Aldershot: Ashgate). And last but not least, the earliest use of 'ETTOing' as a verb is, as far as I can find out, in an email from Captain Arthur Dijkstra, 20 September 2005.

Chapter 1: The Need to Understand Why Things Go Wrong

Safety Requires Knowledge

In the world of risk and safety, in medicine, and in life in general, it is often said that prevention is better than cure. The meaning of this idiom is that it is better to prevent something bad from happening than to deal with it after it has happened. Yet it is a fact of life that perfect prevention is impossible. This realisation has been made famous by the observation that there always is something that can go wrong. Although the anonymous creator of this truism never will be known, it is certain to have been uttered a long time before either Josiah Spode (1733–1797) or the hapless Major Edward A. Murphy Jr. (A more sardonic formulation of this is Ambrose Bierce's definition of an accident as '(a)n inevitable occurrence due to the action of immutable natural laws.')

A more elaborate argument why perfect prevention is impossible was provided by Stanford sociologist Charles Perrow's thesis that socio-technical systems had become so complex that accidents should be considered as normal events. He expressed this in his influential book *Normal Accidents*, published in 1984, in which he presented what is now commonly known as the *Normal Accident Theory*. The theory proposed that many socio-technical systems, such as nuclear power plants, oil refineries, and space missions, by the late 1970s and early 1980s had become so complex that unanticipated interactions of multiple (small) failures were bound to lead to unwanted outcomes, accidents, and disasters. In other words, that accidents should be accepted as normal rather than as exceptional occurrences in such systems. (It was nevertheless soon pointed out that some organisations seemed capable of managing complex and risky environments with higher success rates and fewer accidents than expected. This led to the school of thought known as the study of High Reliability Organisations.)

Even though it is impossible to *prove* that everything that can go wrong will go wrong, everyday experience strongly suggests that this supposition is valid. There is on the other hand little evidence to suggest that there are some things that cannot go wrong. (Pessimists can easily argue that the fact that nothing has never gone wrong or not gone wrong for some time, cannot be used to prove, or even to argue, that it will not go wrong sometime in the future.) Since things that go wrong sometimes may lead to serious adverse outcomes, such as the loss of life, property and/or money, there is an obvious advantage to either prevent something from going wrong or to protect against the consequences. But it is only possible to prevent something from happening if we know *why* it happens, and it is only possible to protect against specific outcomes if we know *what* they are and preferably also *when* they are going to happen. Knowing *why* something happens is necessary to either eliminate or reduce the risks. Knowing *what* the outcomes may be makes it possible to prepare appropriate responses and countermeasures. And knowing *when* something will happen, even approximately, means that a heightened state of readiness can be established when it is needed. Without knowing the *why*, the *what*, and the *when*, it is only possible to predict what may happen on a statistical basis. There is therefore a genuine need to be able better to understand why certain things have happened, or in other words to know how to construct useful and effective explanations.

Despite the universal agreement that safety is important, there is no unequivocal definition of what safety is. Most people, practitioners and researchers alike, may nevertheless accept a definition of safety as 'the freedom from unacceptable risks' as a good starting point. This definition can be operationalised by representing an accident as the combination of an unexpected event and the lack of protection or defence (Figure 1.1). This rendering suggests that safety can be achieved in three different ways: by eliminating the risk, by preventing unexpected events from taking place, and by protecting against unwanted outcomes when they happen anyway.

In most industrial installations risk elimination and prevention are achieved by making risk analysis part of the design and by establishing an effective safety management system for the operation. Protection is achieved by ensuring the capacity to respond to at least the regular threats, and sometimes also the irregular ones.

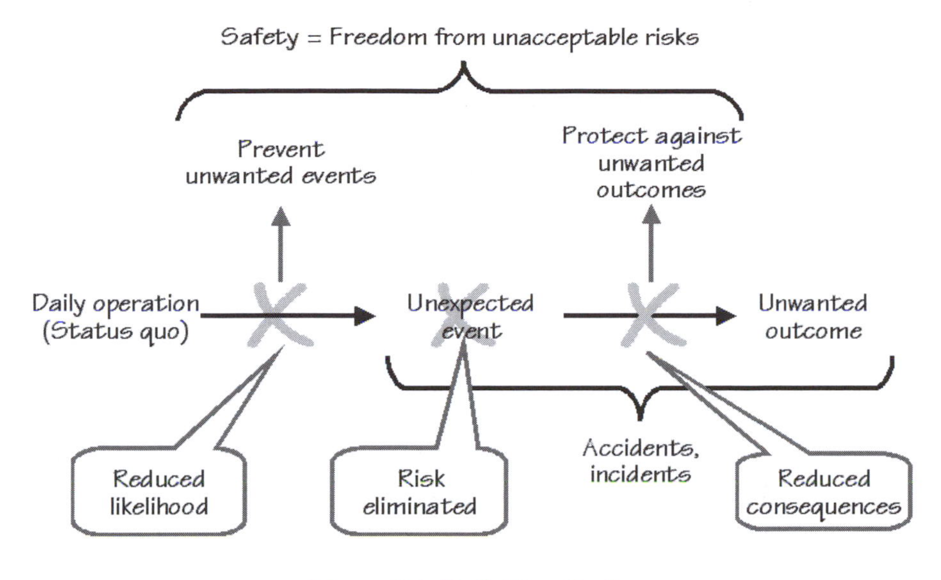

Figure 1.1: Definition of safety

Although the definition of safety given in Figure 1.1 looks simple, it raises a number of significant questions for any safety effort. The first is about what the risks are and how they can be found, i.e., about what can go wrong. The second is about how the freedom from risk can be achieved, i.e., what means are available to prevent unexpected events or to protect against unwanted outcomes. The third has two parts, namely how much risk is acceptable and how much risk is affordable. Answering these questions is rarely easy, but without having at least tried to answer them, efforts to bring about safety are unlikely to be successful.

A Need for Certainty

An acceptable explanation of an accident must fulfil two requirements. It must first of all put our minds at ease by explaining what has happened. This is usually seen as being the same as finding the cause of why something happened. Yet causes are relative rather than absolute and the determination of the cause(s) of an outcome is a social or psychological rather than an objective or technical process. A cause can pragmatically be defined as the identification, after the fact, of a limited set of factors or conditions that provide the necessary and sufficient conditions for the effect(s) to have occurred. (Ambrose Bierce's sarcastic definition of logic as '*The art of thinking and reasoning in strict accordance with the limitations and incapacities of the human misunderstanding*' also comes to mind.) The qualities of a 'good' cause are (1) that it *conforms* to the current norms for explanations, (2) that it can be *associated* unequivocally with a known structure, function, or activity (involving people, components, procedures, etc.) of the system where the accident happened, so (3) that it is *possible* to do something about it with an acceptable or affordable investment of time and effort.

The motivation for finding an explanation is often very pragmatic, for instance that it is important to understand causation as quickly as possible so that something can be done about it. In many practical cases it seems to be more valuable to find an acceptable explanation sooner than to find a meaningful explanation later. (As an example, French President Nicolas Sarkozy announced shortly after a military shooting accident in June 2008, that he expected 'the outcome of investigations as quickly as possible so the exemplary consequences could be taken.') Indeed, whenever a serious accident has happened, there is a strong motivation quickly to find a satisfactory explanation of some kind. This is what the philosopher Friedrich Nietzsche in *The Twilight of the Idols* called 'the error of imaginary cause,' described as follows:

> [to] extract something familiar from something unknown relieves, comforts, and satisfies us, besides giving us a feeling of power. With the unknown, one is confronted with danger, discomfort, and care; the first instinct is to abolish these painful states. First principle: any explanation is better than none. ... A causal explanation is thus contingent on (and aroused by) a feeling of fear.

Although Nietzsche was writing about the follies of his fellow philosophers rather than about accident investigations, his acute characterisation of human weaknesses also applies to the latter. In the search for a cause of an accident we do tend to stop, in the words of Nietzsche, by 'the first interpretation that explains the unknown in familiar terms' and 'to use the feeling of pleasure ... as our criterion for truth.'

Explaining Something that is Out of the Ordinary

On 20 March 1995, members of the Aum Shinrikyo cult carried out an act of domestic terrorism in the Tokyo subway. In five coordinated attacks at the peak of the morning rush hour, the perpetrators released sarin gas on the Chiyoda, Marunouchi, and Hibiya lines of the Tokyo Metro system, killing 12 people, severely injuring 50 and causing temporary vision problems for nearly a thousand others. (Sarin gas is estimated to be 500 times as toxic as cyanide.)

Initial symptoms following exposure to sarin are a runny nose, tightness in the chest and constriction of the pupils, leading to vision problems. Soon after, the victim has difficulty breathing and experiences nausea and drooling. On the day of the attack more than 5,500 people reached hospitals, by ambulances or by other means. Most of the victims were only mildly affected. But they were faced with an unusual situation, in the sense that they had some symptoms which they could not readily explain. The attack was by any measure something extraordinary, but explanations of symptoms were mostly ordinary. One man ascribed the symptoms to the hay-fever remedies he was taking at the time. Another always had headaches and therefore thought that the symptoms were nothing more than that; his vomiting was explained as the effects of 'just a cold.' Yet another explained her nausea and running eyes with a cold remedy she was taking. And so on and so forth.

Such reactions are not very surprising. Whenever we experience a running nose, nausea, or headache, we try to explain this in the simplest possible fashion, for instance as the effects of influenza. Even when the symptoms persist and get worse, we tend to stick to this explanation, since no other is readily available.

The Stop Rule

While some readers may take argument with Nietzsche's view that the main reason for the search for causes is to relieve a sense of anxiety and unease, others may wholeheartedly agree that it often is so, not least in cases of unusual accidents with serious adverse outcomes. Whatever the case may be, the quotation draws attention to the more general problem of the *stop rule*. When embarking on an investigation, or indeed on any kind of analysis, there should always be some criterion for when the analysis stops. In most cases the stop rule is unfortunately only vaguely described or even left implicit. In many cases, an analysis stops when the bottom of a hierarchy or taxonomy is reached. Other stop rules may be that the explanation provides the coveted psychological relief (Nietzsche's criterion), that it identifies a generally acceptable cause (such as 'human error'), that the explanation is politically (or institutionally) convenient, that there is no more time (or manpower, or money) to continue the analysis, that the explanation corresponds to moral or political values (and that a possible alternative would conflict with those), that continuing the search would lead into uncharted – and possibly uncomfortable – territory, etc.

Whatever the stop rule may be and however it may be expressed, an accident investigation can never find the real or true cause simply because that is a meaningless concept. Although various methodologies, such as Root Cause Analysis, may claim the opposite, a closer look soon reveals that such claims are unwarranted. The relativity of Root Cause Analysis is, for instance, obvious from the method's own definition of a root cause as the most basic cause(s) that can reasonably be identified, that management is able to fix, and that when fixed will prevent or significantly reduce the likelihood of the problem's reoccurrence. Instead, a more reasonable position is that an accident investigation is the process of *constructing* a cause rather than of *finding* it. From this perspective, accident investigation is a social and psychological process rather than an objective, technical one.

The role of the stop rule in accident investigation offers a first illustration of the Efficiency-Thoroughness Trade-Off (ETTO) principle. Since the purpose of an accident investigation is to find an adequate explanation for what has happened, the analysis should clearly be as detailed as possible. This means that it should not stop at the first cause it finds, but continue to look for alternative explanations and possible contributing conditions, until no reasonable doubt about the correctness of the outcome remains. The corresponding stop rule could be that the analysis should be continued until it is clear that a continuation will only marginally improve the outcome. To take an analysis beyond the first cause will, however, inevitably take more time and require more resources – more people and more funding. In many cases an investigatory body or authority has a limited number of investigators, and since accidents happen with depressing regularity, only limited time is set aside for each. (Other reasons for not going on for too long are that it may be seen as a lack of leadership and a sign of uncertainty, weakness, or inability to make decisions. In the aftermath of the blackout in the US on 14 August 2003, for example, it soon was realised that it might take many months to establish the cause with certainty. One comment was that 'this raises the possibility that Congress may try to fix the electricity grid before anyone knows what caused it to fail, in keeping with the unwritten credo that the appearance of motion – no matter where – is always better than standing still.') Ending an analysis when a sufficiently good explanation has been found, or using whatever has been found as an explanation when time or resources run out, even knowing that it could have been continued in principle, corresponds to a criterion of efficiency.

We shall call the first alternative *thoroughness*, for reasons that are rather obvious. And we shall call the second alternative *efficiency*, because it produces the desired effect with a minimum of time, expense, effort or waste.

A Need for Simple Explanations

Humans, as individuals or as organisations, prefer simple explanations that point to single and independent factors or causes because it means that a single change or response may be sufficient. (This, of course, also means that the response is cheaper to implement and easier to verify.) An investigation that emphasises efficiency rather than thoroughness, will produce simpler explanations. The best example of that is in technological troubleshooting and in industrial safety. Yet we know that it is risky to make explanations too simple because most things that happen usually – if not always – have a complex background.

Another reason for preferring simple explanations is that finding out what has happened – and preparing an appropriate response – takes time. The more detailed the explanation is, the longer time it will take. If time is too short, i.e., if something new happens before there has been time to figure out what happened just before – and even worse, before there has been time to find a proper response – the system will run behind events and will sooner or later lose control. In order to avoid that it is necessary to make sure that there is time to spare. Explanations should therefore be found fast, hence be simple. The sustained existence of a system depends on a trade-off between efficiency – doing things, carrying out actions before it is too late – and thoroughness – making sure that the situation is correctly understood and that the actions are appropriate for the purpose.

The trade-off can be illustrated by a pair of scales (Figure 1.2), where one side represents efficiency and the other thoroughness. If efficiency dominates, control may be lost because actions are carried out before the conditions are right or because they are the wrong actions. If thoroughness dominates, actions may miss their time–window, hence come too late. In both cases failure is more likely than success. In order for performance to succeed and for control to be maintained, efficiency and thoroughness must therefore be in balance.

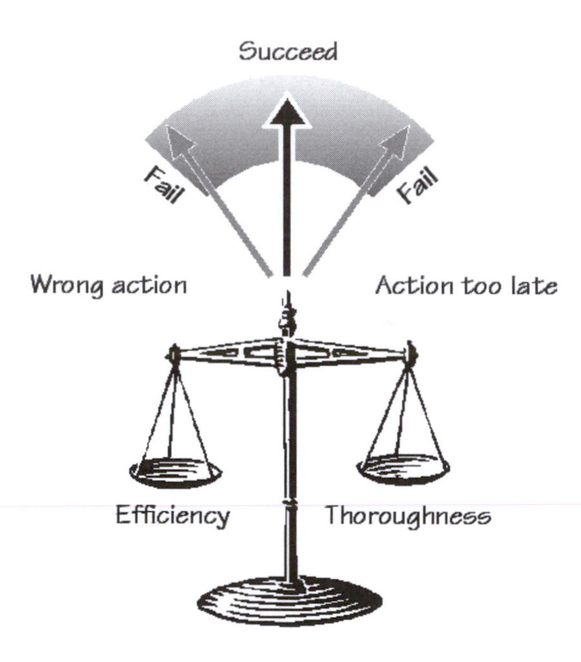

Figure 1.2: Balance between efficiency and thoroughness

Humans, as individuals or as organisations, can with practice learn just how much time they can spend on deliberating before acting, so that the chance of success is sufficiently high. This can either mean that efficiency is sacrificed for thoroughness, or – more often – that thoroughness is sacrificed for efficiency. One way of saving time and efforts is to make shortcuts in reasoning. However, the shortcuts also introduce some uncertainty and increase the unpredictability of the situation. If the unpredictability becomes too large, actions are more likely to fail than to succeed. In order to manage safety we therefore need to know how this trade-off is done – realising, of course, that the same principle applies to our analysis and description of it.

The ETTO Principle

The essence of the balance or trade-off between efficiency and thoroughness can be described by the ETTO principle. In its simplest possible form, it can be stated as follows: in their daily activities, at work or at leisure, people routinely make a choice between being effective and being thorough, since it rarely is possible to be both at the same time. If demands for productivity or performance are high, thoroughness is reduced until the productivity goals are met. If demands for safety are high, efficiency is reduced until the safety goals are met.

The fundamentals of the ETTO principle can be illustrated by the following example, taken from Herbert William Heinrich's pioneering work, *Industrial Accident Prevention,* published in 1931:

> In splitting a board, a circular-saw operator suffered the loss of his thumb when, in violation of instructions, he pushed the board past the saw with his fingers, instead of using the push stick that had been provided for the purpose. He stated that he had always done such work in this manner and had never before been hurt. He had performed similar operations on an average of twenty times a day for three months and had therefore exposed his hand in this way over one thousand five hundred times.

In this case thoroughness is represented by the instructions, which mandated the use of the push stick, a hand-held device used to push the board past the saw while keeping hands some distance from the cutter. The benefit is that if something goes wrong, then the push stick rather than the hands will be damaged. While using the push stick is safe, it is also less efficient than using the hands to push the board. Using the hands eliminates two or three steps (positioning the board on the saw-table, grabbing the push stick, positioning the push stick relative to the board) hence saves both time and effort. For a recurrent work situation most people will naturally choose the more efficient mode of operation as long as it, in their experience, is just as safe as the alternative.

The ETTO principle has been introduced using accidents and accident investigations as examples. But its application is much wider, and it can be proposed as a common principle to describe both individual and organisational performance. The ETTO principle can be found in all kinds of human activity, both as a characteristic of the explanations that we use and as a characteristic of the way in which we find the explanations. The ETTO principle can be used to understand how things are done, and why they generally succeed but sometimes fail. It can therefore also be used better to predict what may happen – both in the sense of successes and in the sense of failures.

A Definition of Efficiency. Efficiency means that the level of investment or amount of resources used or needed to achieve a stated goal or objective are kept as low as possible. The resources may be expressed in terms of time, materials, money, psychological effort (workload), physical effort (fatigue), manpower (number of people), etc. The appropriate level or amount is determined by the subjective evaluation of what is sufficient to achieve the goal, i.e., good enough to be acceptable by whatever stop rule is applied as well as by external requirements and demands. For individuals, the decision about how much effort to spend is usually not conscious, but rather a result of habit, social norms, and established practice. For organisations, it is more likely to be the result of a direct consideration – although this choice in itself will also be subject to the ETTO principle.

A Definition of Thoroughness. Thoroughness means that an activity is carried out only if the individual or organisation is confident that the necessary and sufficient conditions for it exist so that the activity will achieve its objective and not create any unwanted side-effects. More formally, thoroughness means that the preconditions for an activity are in place, that the execution conditions can be ensured, and that the outcome(s) will be the intended one(s).

A simple example of preconditions is the check-lists that pilots go through before taxiing to the runway. An example of execution conditions is, for instance, making certain that there is enough blood for a patient during an operation, or enough petrol in the tank for a car to reach its destination. And finally, an example of ensuring the proper outcome is the checking of drugs before they are approved for release on the market (cf. Chapter 7) – or more generally, the use of the precautionary principle.

The ETTO Principle in Practice

It is clearly necessary for any individual or organisation to be both efficient and thorough. It is necessary to be efficient because resources always are limited, and in particular because time is limited. (Time can of course be considered as just another resource, but its special character warrants that it is treated separately.) It is likewise necessary to be thorough both to make sure that we do things in the right way, so that we can achieve what we intend, and to avoid adverse consequences – incidents and accidents. One simple reason why thoroughness is needed is that accidents and incidents invariably will reduce the efficiency, for instance because the system has to be stopped for shorter or longer periods of time, parts repaired or replaced, verified for use again, etc. More concretely, an efficiency-thoroughness trade-off can happen for one or more of the following reasons:

- Limited availability of required resources, especially limited time or uncertainty about how much time is available.
- The natural tendency – or indeed human propensity – not to use more effort than needed.
- A need – often implicit or unstated – to maintain a reserve (of resources, of time) in case something unexpected happens.
- Social pressures from managers, colleagues, or subordinates, for instance to do things in a certain way or by a certain time.
- Organisational pressures, for instance a conflict between official priorities ('*safety first*') and actual practices ('*be ready in time*'), or a lack of resources.
- Individual priorities, habits of work, ambition, etc.

The efficiency-thoroughness trade-off is a characteristic of people, because they are flexible and intelligent beings, and therefore also of organisations as socio-technical systems. The efficiency-thoroughness trade-off cannot be made by machines unless it is part of their design or programming. Yet in such cases the trade-off is different from what humans do, since it is algorithmic rather than heuristic. Indeed, it can be argued that the performance of any living system that has a modicum of awareness of its own existence will show this trade-off in one way or the other. An example from the world of birds will be given in Chapter 5.

Although the ETTO principle is useful to describe a pervasive functional characteristic of individuals and organisations, it is not put forward as an explanation in the sense of a 'true' cause of what happens. It can be useful as a descriptive principle inasmuch as it allows us to understand better what people do – and therefore in a sense also why they do it – and therefore to predict how humans and organisations will behave in certain situations. In that sense it allows us to be more effective, hence vindicates itself.

Changing Views

It is a firmly entrenched practice to assume that adverse outcomes must be explained by failures and malfunctions of system components. While this may be a valid position vis-à-vis technological systems and machines, it is not valid for humans, organisations, and socio-technical systems. Instead of seeing successes and failures as two separate categories of outcomes brought about by different processes or 'mechanisms,' we must agree with Resilience Engineering that they are but two sides of the same coin. As the physicist and philosopher Ernst Mach noted in 1905 'knowledge and error flow from the same mental sources, only success can tell one from the other.' We should therefore try to understand the sources of performance rather than just the sources of failure. To extend the scope from individual to organisational performance, we may paraphrase Mach to say that *success and failure have the same origins, only the outcome can distinguish one from the other.*

Things go wrong for the very same reasons that things go right. This does not mean that the 'incorrect' actions that can explain an accident are right, but rather that the way in which these actions were made – and therefore also the principles by which they must be understood – was the same regardless of whether the outcome was a success or a failure. It is just the momentary variability or juxtaposition of conditions that led to the wrong or unwanted result. This book is about understanding those principles. And paradoxically, as shall be shown at the end, understanding those principles can to some extent free us from them.

A Note on Terminology

This chapter has referred to something called a 'system' or a 'socio-technical system.' A system can broadly be defined as the intentional organisation or arrangement of parts (components, people, functions, subsystems) that makes it possible to achieve specified and required goals. There are two important aspects of this definition. Concerning the parts that make up the system, we shall exclusively be referring to socio-technical systems. The idea of a socio-technical system is that the conditions for successful organisational performance – and conversely also for unsuccessful performance – are created by the interaction between social and technical factors. The other aspect concerns the boundary of the socio-technical system. While it is common to define the boundary relative to some structure or physical characteristic, we shall define the boundary based on how the system functions, in accordance with the principles of cognitive systems engineering. The boundary is therefore always relative to the purpose or scope of the description. This makes sense since the ETTO principle refers to the system's dynamics rather than to its architecture.

Sources for Chapter 1

The original formulation of Murphy's law is, allegedly, 'If there's more than one way to do a job, and one of those ways will result in disaster, then somebody will do it that way.' The Wikipedia entry on Murphy's law describes both its history and several alternative versions.

The *Normal Accident Theory* has already been mentioned in the sources for the Prologue. It has been described in the following work: C. Perrow, (1984), *Normal Accidents: Living with High-risk Technologies* (New York, USA: Basic Books, Inc). Soon after it was published, researchers from both UC Berkeley and the University of Michigan pointed out that there were some organisations that apparently managed to avoid accidents even though they existed in risky and complex environments. These organisations were dubbed *High Reliability Organisations* (*HRO*). A good introduction to this line of work is K. H. Roberts, (ed.) (1993), *New Challenges to Understanding Organizations,* (New York: Macmillan).

More details on the definition of safety can be found in E. Hollnagel, (2004), *Barriers and Accident Prevention* (Aldershot: Ashgate).

The International Civil Aviation Organisation (ICAO), for instance, defines safety as '... the state in which the risk of harm to persons or of property damage is reduced to, and maintained at or below, an acceptable level through a continuing process of hazard identification and risk management.' Safety normally describes the state or condition of being safe (derived from the Latin word *salvus*), i.e., being protected against various types of consequences of non-desirable events. In recent years it has been proposed by Resilience Engineering that safety should be seen as a process rather than a state, i.e., as something a system *does* rather than something a system *has*.

The definitions of Ambrose Bierce are taken from *The Devil's Dictionary* which was published in 1911. An on-line version of this delightful work is available at http://www.thedevilsdictionary.com/

The Tokyo gas attack has been described in Haruki Murakami's very interesting book *Underground* (Vintage International, 2000). The book is based on extensive interviews with both victims and perpetrators. Murakami is otherwise best known for his excellent fiction.

A fundamental discussion of accident investigation as a social process is provided by D. D. Woods, L. J. Johannesen, R. I. Cook and N. B. Sarter, (1994), *Behind Human Error: Cognitive Systems, Computers, and Hindsight* (Wright-Patterson Air Force Base, OH: Crew System Ergonomics Information Analysis Center.)

The example of the circular-saw operator is found on page 94 of H. W. Heinrich (1931), *Industrial Accident Prevention* (New York: McGraw-Hill). This book pioneered the practical study of industrial safety, and introduced a number of models and methods, several of which are still in use. Mr Heinrich was an Assistant Superintendent of the Engineering and Inspection Division of Travelers Insurance Company in Hartford, CT, and the book was based on his extensive practical experience in accident investigation.

The definition of what a system is can be a touchy issue and is often neglected. An alternative to the definition used here is to define a system as 'a set of objects together with relationships between the objects and between their attributes' (D. A. Hall and R. E. Fagen (1968), Definition of System. In W. Buckley (ed), *Modern Systems Research for the Behavioural Scientist* (Chicago: Aldine Publishing Company), p. 81). Other definitions can be found in the general cybernetics literature. A good reference for readers who would like to know more is

P. Checkland (2000), *Systems Thinking, Systems Practice* (Chichester, UK: Wiley). The problem of how to define boundaries is also discussed in E. Hollnagel and D. D. Woods (2005), *Joint Cognitive Systems: Foundations of Cognitive Systems Engineering* (Boca Raton, FL: Taylor & Francis).

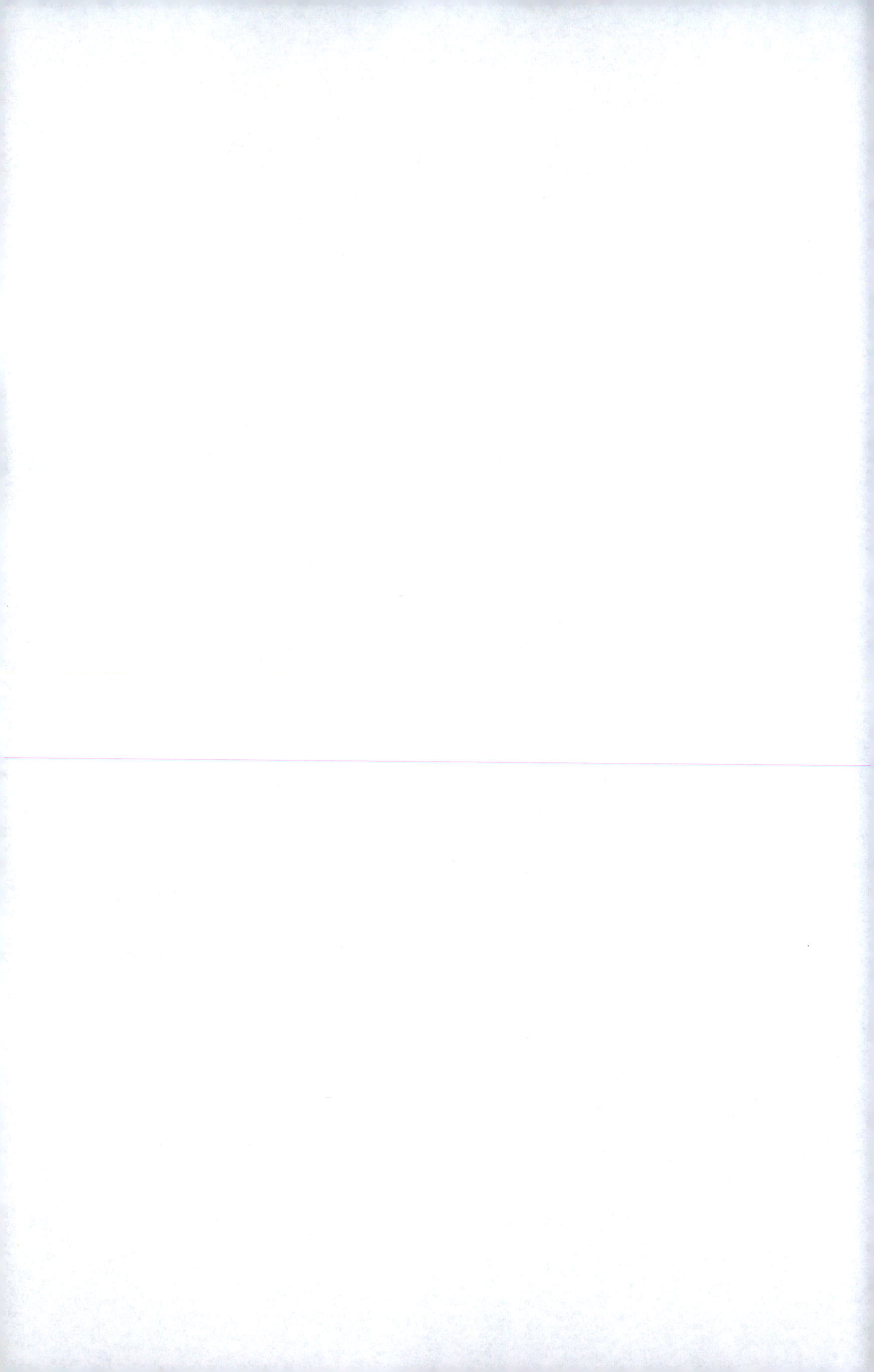

Chapter 2: From Rationality to ETTOing

In the scientific disciplines that try to account for human behaviour and performance, be it with regard to morals, decision-making, the design of human–machine interaction, etc., there is no shortage of advice or recommendations for how people should behave and what they should do to make sure that the desired outcomes are achieved. The common assumption is that if people follow the rules or the instructions, i.e., if they are rational, then they will always succeed. Conversely, if they do not, they are likely to fail. And with a little bit of incorrect reasoning – itself against the rules of logic – it follows that if people fail, then they must have broken the rules.

The origin of this thinking goes far back in time, ranging from millennia in the case of religion and philosophy, to centuries in the case of gambling and work. The classical description of the rational human is expressed by the theory of decision-making, especially when this is combined with economics. Classical or normative decision theory, and specifically the part of decision theory that deals with economic behaviour, often refers to an ideal decision-maker know as *homo economicus* or the rational economic (hu)man.

The Rational Human

The origins of the *homo economicus* can be traced back, at least to the British philosopher Jeremy Bentham (1738–1832), and probably even to the early attempts to calculate probabilities for outcomes of games. A game is a situation where a player has to make decisions about future outcomes, typically about whether or not to place a bet. If the decision is right the player wins, but if the decision is wrong the player loses. In order make the right decision it is necessary to know all the possible outcomes, the probability of each outcome, and the value of each outcome. For games it is relatively easy to obtain this information, not least because the outcome can be equated with the economic value of the win (or loss). For decision-making in the face of the vagaries of working life, the situation is more difficult. A *homo economicus* must more specifically meet the following requirements:

- Be *completely informed.* This means that the decision-maker knows what all the possible alternatives are, and also knows what the outcome of any action will be. This implies that the decision-maker's beliefs are based on a logical, objective analysis of all the available evidence.
- Be *infinitely sensitive.* It is obviously necessary that the decision-maker can discriminate among alternatives. This presupposes that the decision-maker is able to perceive any difference that may exist, no matter how small it is. The decision-maker must therefore – in principle – be infinitely sensitive.
- Be *rational*, which means that the decision-maker is able to rank the alternatives according to some criterion, such as utility, and choose so as to maximise something. Thus if the choice is between *Wein, Weib, und Gesang*, the rational decision-maker must be able to put them in an order of preference and to make a choice based on that.

When it comes to the criterion, decision theory has historically made several suggestions for what should be maximised. Examples include John Stuart Mill's 'general happiness,' subjective expected utility, risk avoidance, loss aversion, safety, pleasure, etc. The problem with proposing a criterion is that once a decision has been made, then it is always possible to claim *post hoc* that something was maximised. (In practice, people may succumb to *post-decision consolidation*, to justify the choice they made.) Formal decision theory, however, applies the principle of maximisation in a stronger sense, namely that the decision-maker must use a recognisable criterion to choose among the alternatives. Unless that is the case, it becomes impossible to make 'rational' decisions – and this is after all what decision theory is about.

The problem with *homo economicus* is that there is a significant difference between what people should do and what they actually do. There are several good reasons for this. First, even if the criteria are relative rather than absolute (with regard to information, sensitivity, and ordering), rational decision-making still requires a colossal level of consideration and thinking, far beyond what an unaided human is able to provide. Second, classical decision theory assumes that problems can be addressed and solved one by one. But in practice it is often difficult to separate problems and they must instead be looked at as a whole. Third, it takes time to make a decision and knowledge about alternatives and outcomes may become outdated or degrade as time goes by. In practice, people do not behave as rational decision-makers or logical information processors, but rather try to get by in other ways.

The rational ideal received a welcome boost of support in the 1950s when it was realised that the brain could be described as a digital computer and that humans consequently could be described as information processors. Since a computer was a rational machine, and since it suddenly seemed possible to describe humans as computers, the optimism of *homo economicus* was revived. Despite the widespread enthusiasm, a number of studies in the late 1960s and early 1970s made clear that humans as decision-makers used a number of strategies or heuristics that enabled them to cope with the complexity of the situations even though the strategies were far from rational in the normative sense of the term. This posed a problem (and still does) for decision theory and psychology. The researchers tried to solve this with considerable ingenuity by proposing alternative theories that on the one hand better matched what people did in practice, but on the other retained some kind of formal principle and required that humans behaved or performed in accordance with this principle. The rational decision-making perspective was thereby replaced by various forms of limited or bounded rationality, which in turn rested on assumptions of limited capabilities of memory, attention, etc. Examples of this will be provided in Chapter 3.

Time to Think, and Time to Do

The predominant explanations from psychologists and engineers attribute performance failures to the mismatch between demand and capacity (something that is also characteristic of the rational decision-making perspective), but curiously neglects the fundamental fact that everything takes time and takes place in time.

For many theories, such as classical economic decision theory, it is convenient to assume that alternatives as well as criteria are constant while the decision is made, that hence time does not exist. But this assumption is clearly not tenable except under very unusual circumstances. If, on the other hand, it is acknowledged that time is important, then there is less need of elaborate theories of bounded rationality or information capacity limits. Limited capacity is not a quality of the system in isolation, but describes the inability to be fast enough – or effective enough – for the time that is available. That goes for daily activities at work as well as an *oeuvre du vie* – and, needless to say, for writing a book such as this.

Most descriptions of decision-making divide it into a number of steps. Exactly how many and what they are called depend on the model in question. But there is general agreement that it is necessary both to orient oneself and evaluate the situation, decide what to do and plan how to do it, and then finally carry it out. We can for simplicity's sake call these steps evaluation, selection, and execution. The steps are often represented as a set of boxes linked by non-descript arrows, cf. Figure 2.1. To the extent that each step has a duration, these durations are simply added on the assumption that the steps follow one another in a sequence. Apart from that, time is not considered at all.

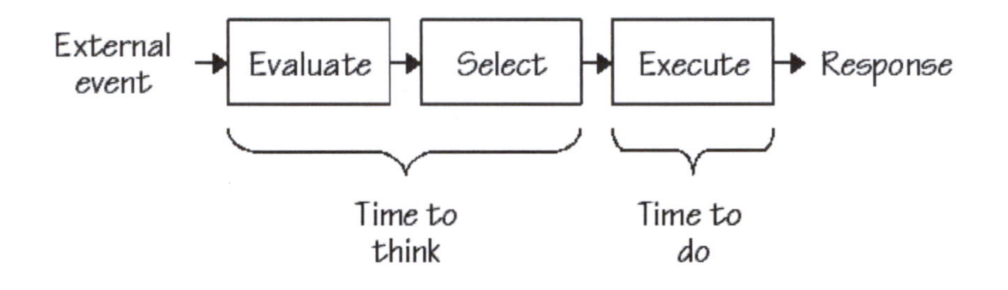

Figure 2.1: Sequential representation of action steps

Figure 2.2 shows a different interpretation. In this case the three stages can overlap, indicating that more or less time can be allocated to each. If each stage was allowed to last as long as needed, this would define the maximum time required. But the stages usually overlap. (Figure 2.2 is an idealisation in the sense that the stages in reality may iterate rather than follow each other in a sequence.) In addition to the time needed for each stage, there is also the time that is available to make the decision and carry out the action. As Figure 2.2 shows, the time available and the time needed are usually not in complete agreement, and very often more time is needed than is available. The 'solution' is, of course, to reduce the time either for evaluation, for selection, or for execution, so that the time required does not exceed the time available.

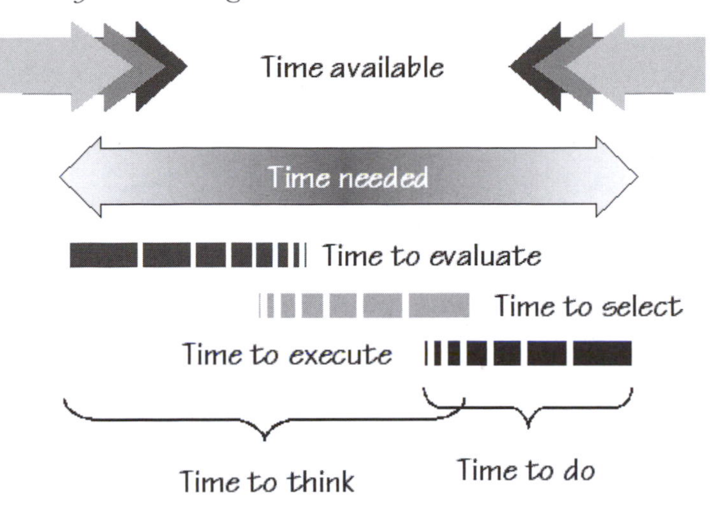

Figure 2.2: Overlapping representation of action steps

A further complication, not shown in Figure 2.2, is that the time when an activity must be completed often is unknown. It may, of course, be possible to estimate how long it will take to do something, hence the earliest time it can be completed. But the time available is sometimes uncertain because the latest finishing time is unknown. Another factor is that unexpected events may demand attention, so that the current activity is completed prematurely or suspended.

One way of representing this is shown in Figure 2.3. Here the three steps have been arranged in a cycle to emphasise the continuous nature of human actions. The external event can be the start of a cycle, but also the end of it, in the sense that a new external event may require attention and therefore interrupt an ongoing activity. Since the decision-maker, even in well-organised environments, cannot know with certainty when the next event will happen, it makes sense to play it safe and finish as soon as possible, since this may leave time to spare in case something unexpected should happen.

Whenever an acting organism – such as an individual, a team, or an organisation – must engage in an activity, there are always two options. One is to wait, to gather more information, to see how things develop, or just to hope for a greater level of certainty – or less uncertainty. The other is to go ahead on the assumption that the situation is known well enough and the alternatives are clear enough – and indeed that all reasonable alternatives are known. It is this dilemma between time to think and time to do that is at the heart of the ETTO principle.

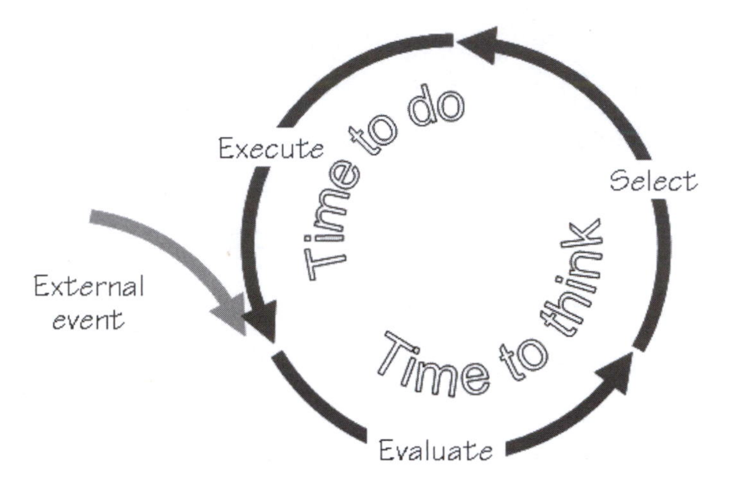

Figure 2.3: Cyclical representation of action steps

ETTO Redefined

The available time will in practice always be limited and usually too short. Since the mismatch between the time needed to do something and the time available only rarely can be resolved by increasing the time available, such as in renegotiating deadlines and delivery schedules, the only remaining option is to reduce the time needed, for instance by working faster or by simplifying the tasks. Every activity requires that a certain minimum of information is available, lest it be reduced to trial-and-error. Every activity also requires a certain minimum effectiveness of action, lest it is overtaken by events. The ETTO principle describes the way in which people (and organisations) ensure that the minima are attained, and that the conditions are improved as far as possible.

The definition of the ETTO principle from Chapter 1 can be rewritten as follows: The ETTO principle refers to the fact that people (and organisations) as part of their activities frequently – or always – have to make a trade-off between the resources (time and effort) they spend on preparing an activity and the resources (time and effort) they spend on doing it. The trade-off may favour thoroughness over efficiency if safety and quality are the dominant concerns, and efficiency over thoroughness if throughput and output are the dominant concerns. It follows from the ETTO principle that it is never possible to maximise efficiency and thoroughness at the same time. Nor can an activity expect to succeed, if there is not a minimum of either.

Feed the Birds

A bird in the wild, particularly if it is small, is constantly exposed to the conflict between foraging, finding something to eat, and vigilance, avoiding being caught by a predator. Foraging normally requires a head-down position, from which it is difficult if not impossible to detect a possible predator. Vigilance requires a head-up position, which makes it impossible to find and pick up any food. If a bird therefore tries to maximise its intake of energy, it increases the risk of being caught by a predator. Yet if it tries to minimise the risk from predators, it also reduces the intake of energy. The partly unpredictable environment of wild birds therefore requires them to find a balance or a strategy that increases their chances of survival on both counts.

Figure 2.4: Foraging bird
(c) Digitaldave | Dreamstime.com

Birds obviously do not reason about this in the same way that humans do, but their behaviour nevertheless expresses the ETTO principle, efficiency being foraging and thoroughness being vigilance. Studies have furthermore shown that this trade-off is affected by the conditions, rather than being a built-in, fixed ratio. For instance, before migratory birds begin their voyage, the foraging–vigilance ratio changes to increase the proportion of foraging and reduce the proportion of anti-predatory behaviour. This makes sense (from a human interpretation, of course) since it is more important to accumulate resources before a long flight.

Grazing animals in the wild, e.g., zebras and buffaloes, display the same behaviour, but domesticated animals do not. The reason is that they have learned that there are no predators, hence no need to be vigilant. A cow, for instance, spends far more head-down time than a gazelle. The human analogy may be that when we know that there is no danger, or rather when we lose our sense of unease, then efficiency takes over and thoroughness goes out the door. As long as the assumption is right and there is no risk, it is safe. But as soon as there is a risk, it is not. And human – and organisational – memory is unfortunately rather short.

The Value of Planning

The Roman army was probably the best organised fighting force that has ever been. It was well structured, not only in the way it was organised – the famous legions – but also in its daily operations. One hallmark of that was the Roman army camp, whether it was a permanent or a temporary one. As regards the latter, a legion would build a *castrum* at the end of a day's march. It was built by specialised engineering units and would often be ready after a few hours. This was in itself an impressive feat of engineering, but more important for the present discussion is the fact that every *castrum* throughout the army was built according to the same layout (Figure 2.5).

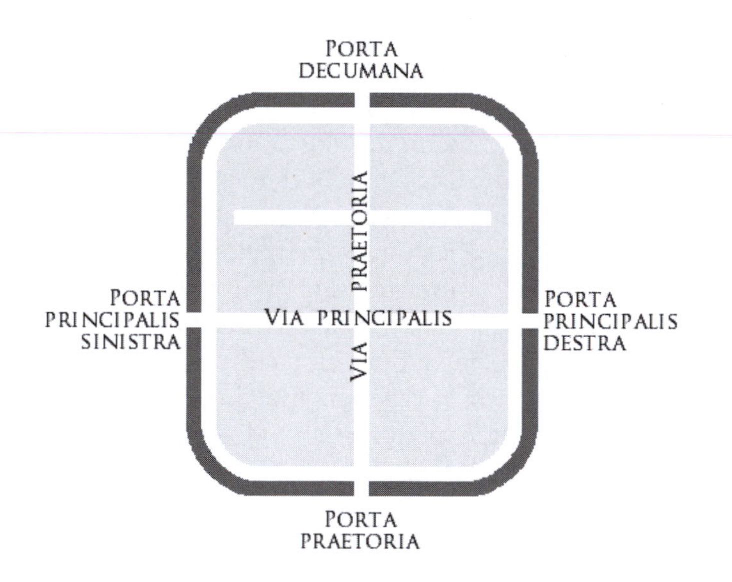

Figure 2.5: Standard layout of a Roman army camp (castrum)

Using the same layout obviously makes the construction of the camp more efficient, since fewer questions have to be asked and fewer decisions have to be made. But an equally important reason is that a soldier always would know where to find everything. This was useful if a messenger from one camp entered another, but even more important in the case of an attack during the night. Knowing the layout and where everything was would be an advantage over the enemy, and also an advantage in itself, if soldiers were suddenly aroused from their sleep. In this case the thoroughness that was achieved through preparation, planning, and routine would improve the efficiency during operation, especially during a battle. In terms of the concepts introduce above, the time to evaluate and time to select would be reduced, leaving more – and hopefully sufficient – time to act.

Information Pull and Information Push

The development and rapid commercialisation of wireless networks has made it possible to provide information about (almost) anything, at (almost) any time, and (almost) anywhere. Skilful marketeers have even managed to convince many of us that this is a desirable state of affairs. The use of this technology can be described as a situation of either information pull or information push. Information pull is when a user or consumer takes the initiative to get information, i.e., 'pulls' it from the source. Going to the library is an example of information pull and so is searching on the internet. Information push is when the information provider or supplier takes the initiative to deliver it, i.e., that it is 'pushed' upon the consumer or user.

Even prior to the availability of the wireless internet, information push was part of many industrial working environments, namely in the form of alarms. In order to manage a dynamic process or development, whether an isomerisation unit at an oil refinery or the stock market, there must be a balance between information pull and information push. In information pull, the user or operator decides when to get information, and will therefore also usually have the time and resources (and readiness) to make use of it. In information push, the information is forced upon the operator who may not be ready to deal with it. In terms of the situation represented by Figure 2.3, information push happens when ongoing activities are interrupted by a new event. If this happens often, the typical response is to be efficient rather than to be thorough, since this reduces the likelihood of losing control.

Consider, for instance, a situation where an operator must keep track of how a process develops – such as a unit in a refinery or the stock market. If there are *n* measurements to keep track of, the operator must develop some kind of scanning strategy that makes it possible regularly to check each measurement. That in effect means scheduling *n* smaller tasks in some way, and also making sure that no single task takes too long or requires too many resources.

Keeping track of the development of a process is demanding and it therefore makes sense to reduce such demands by automating the scanning/monitoring activities. This indeed becomes necessary when the number of measurements grows so large that it is impossible in practice for a human to handle. (That number is probably very small, and often taken to be the 'magical' number seven alluded to in the Prologue.) But automating the condition monitoring changes the situation from information pull to information push. The positive side is that resources become available to do other things, such as keeping an eye on the larger picture, evaluating or re-evaluating past events, planning and scheduling what to do next, or even taking care of an unrelated task (calling someone, making coffee, etc.) The negative side is that the operator no longer knows when the next alarm will occur, hence no longer knows how much time is available to do something. Information push therefore puts a premium on efficiency and on completing each activity as quickly as possible in order to be ready in case something unexpected happens.

Descriptive Decision Rules

The common response to the finding that people are not rational decision-makers has been to modify or weaken the prescriptions in the hope of reducing the difference between the prescribed and the real, rather than to relinquish the idea of prescribing what people should do. This has been done by changing the decision requirements from rationality to optimality, from optimising to maximising, and from maximising to satisficing. What has rarely been done, at least explicitly, is to acknowledge the importance of time and that decisions are not just a choice among alternatives to find the best – however that is defined – but rather a question of making trade-offs or sacrifices.

Chapter 3 will describe three well known approaches in some detail ('muddling through,' *satisficing*, and naturalistic decision-making). For now, the common response can be illustrated by considering a mathematically elegant solution called 'Elimination By Aspects' or EBA, developed by the mathematical psychologist Amos Tversky (1937–1996). The 'Elimination By Aspects' decision principle describes how the selection of the best alternative (in this case a product, describable on several different attributes), takes place by combining two simpler principles: first to go through the attributes in order of descending importance, and second to eliminate the choices attribute by attribute, until only one possibility remains.

Although the EBA principle relaxes the requirements to rationality by relinquishing maximisation, it still demands that the decision-maker is consistent, hence that the process of gradual selection is 'rational.' Note also that this principle does not reduce the efforts needed to reach the final choice, but rather distributes them over time. The EBA can be seen as a proposal for how to maintain thoroughness (in the new, limited sense), although this can only be achieved by reducing efficiency (i.e., the selection process will take longer). The strategy is therefore only feasible if there is time enough.

A related proposal is *prospect theory*, which describes what people should do in situations where they must choose between risky alternatives. This theory presents the decision processes as two stages called editing and evaluation. In the editing stage, possible outcomes of the decision are ordered, using some heuristic, relative to a reference point, from where lower outcomes are considered as losses and larger as gains. In the evaluation stage, people choose the alternative that has the best prospect, based on the potential outcomes and their respective probabilities. The principle is also here that the overall decision process is broken into smaller and hopefully more manageable stages.

Judgement Heuristics

The EBA principle and prospect theory try to maintain rationality by proposing a workable solution. Other studies have acknowledged that people normally rely on various heuristics to reduce the complexity of their task in order to achieve their goals, and that these heuristics generally violate the principle of rationality. Some typical examples are:

- In the recognition of situations, events, objects, etc., the two main heuristics are *similarity matching* and *frequency gambling*. As the names imply, if something (an event, an object) looks similar to or 'like' something already known, it is judged to be the same. Likewise, if something has happened frequently it is deemed more likely to happen again: the person makes a gamble on frequency in the recognition process.
- Judgement of uncertainty. When a choice or a decision must be made, it is necessary to be reasonably certain about what the alternatives are and what they may lead to. Humans often find it difficult to judge the uncertainty and therefore resort to some short cut or heuristic. Typical examples are *representativeness* where commonality between objects of similar appearance is assumed, *availability* where people base their judgement on how easily an example can be brought to mind, and finally *adjustment* and *anchoring*, where people make estimates by starting from an initial value that biases the estimates.
- Concept formation strategies, as in trying to find a rule or a principle that explains a series of events or how something develops. Common heuristics here are: *focus gambling* – opportunistically changing from one hypothesis to another; *conservative focusing* – moving slowly and incrementally to build up a hypothesis; and *simultaneous scanning* – trying out several hypotheses at the same time.
- Finally, decision-making strategies such as *satisficing*, *muddling through*, and *recognition primed decisions*, that will be detailed in Chapter 3.

These heuristics all serve to improve efficiency while preserving a modicum of thoroughness. This is acceptable because it is better in most cases to produce an imperfect reply than to go on deliberating and thereby possibly miss the window of opportunity. The heuristics are often used unintentionally, i.e., they have become second nature so that people rely on them and switch between them if results are not forthcoming quickly enough.

Work Related ETTO Rules

The sections above have already provided several illustrations of the ETTO principle, and argued that it ranges from birds and animals over individuals to companies, organisations and governments. This section and the two following will describe a set of ETTO rules that are commonly found in practice – at workplaces of all kinds, from the factory floor to the board room or government office. The rules are derived from general experience and informal observations shared by colleagues and gleaned from the literature, as well as on the categories embedded in accident investigation methods and found in various psychological theories. In most cases the ETTO rules are self-explanatory, as rules of thumb or institutionalised practices. In the few cases where this is not the case, some words of explanation have been added. The goal has been to provide a set of characteristic or representative rules, rather than a complete list. The latter is probably impossible anyway. Most of these ETTO rules will be exemplified throughout the book, in particular by the case studies presented in Chapter 4.

- 'It looks fine' – so there is no need to do anything, meaning that an action or an effort can safely be skipped.
- 'It is not really important' – meaning that there is *really* no need to do anything now, provided you understand the situation correctly.
- 'It is normally OK, there is no need to check' – it may look suspicious, but do not worry, it always works out in the end. A variation of this is 'I/we have done this millions of times before' – so trust me/us to do the right thing.
- 'It is good enough for now (or for "government work")' – meaning that it passes someone's minimal requirements.
- 'It will be checked later by someone else' – so we can skip this test now and save some time.
- 'It has been checked earlier by someone else' – so we can skip this test now and save some time. A combination of this rule and the preceding is clearly unhealthy, since it opens a path to failure. It happens every now and then, usually because different people are involved at different times.
- '(Doing it) this way is much quicker' – or more resource efficient – even though it does not follow the procedures in every detail.
- 'There is no time (or no resources) to do it now' – so we postpone it for later and continue with something else instead. The obvious risk is, of course, that we forget whatever we postpone.

- 'We must not use too much of X' – so try to find another way of getting it done. (X can be any kind of resource, including time and money.)
- 'I cannot remember how to do it' (and I cannot be bothered to look it up) – but this looks like a reasonable way of going about it.
- 'We always do it in this way here' – so don't be worried that the procedures say something else.
- 'It looks like a Y, so it probably is a Y' – this is a variety of the representativeness heuristic.
- 'It normally works' (or it has worked before) – so it will probably also work now. This eliminates the effort needed to consider the situation in detail in order to find out what to do.
- 'We must get this done' (before someone else beats us to it or before time runs out) – therefore we cannot afford to follow the procedures (or rules and regulations) in every detail.
- 'It must be ready in time' – so let's get on with it. (The need to meet a deadline may be that of the company, of the bosses, or of oneself.)
- 'If you don't say anything, I won't either' – in this situation one person has typically 'bent the rules' in order to make life easier for another person or to offer some kind of service. This trade-off involves more than one person, and is therefore social rather than individual.
- 'I am not an expert on this, so I will let you decide.' This is another kind of social ETTO rule, where time and effort is saved by deferring to the knowledge and experience of another person. This rule applies not only to situations at work, but to many other type of relations, not least consultation of various kinds. In view of the momentuous events in 2008, it might also be called the financial ETTO rule.

Most readers can undoubtedly provide concrete examples for several or all of the ETTO rules based on their own experience – or from what they have seen their colleagues do! In the remaining chapters, and particularly in Chapter 4, we will look in a little more detail at a variety of real-life examples, and characterise them in terms of which ETTO rule, or rules, they used.

Individual (Psychological) ETTO Rules

In addition to the work related ETTO rules, people also use ETTO rules to manage their own situation, e.g., in terms of workload or task difficulty. Such rules can be found for situations of information input overload, general ways of thinking and reasoning (cognitive style), as well as the judgement heuristics already mentioned.

It is practically a defining characteristic of life and work in the industrialised societies that there is too much information available (cf. the discussion of *information pull and push* above). In the behavioural sciences this is known as a situation of information input overload. A number of heuristics for this situation have been described, such as omission, reduced precision, queueing, and filtering; more detail will be provided in Chapter 3.

Cognitive styles refer to an individual's preferred way of thinking, remembering or problem solving. Some characteristic cognitive styles are:

- *Scanning styles* – differences in the way in which assumptions are tested, either by conservative focusing where only one aspect is changed at a time or by focus gambling where more than one attribute is changed at a time.
- *Levelling versus sharpening* – individual variations in the distinctiveness of memories and the tendency to merge similar events.
- *Reflection versus impulsiveness* – differences in the ways in which alternative hypotheses are formed and responses made.
- *Learning strategies* – a holist gathers information randomly within a framework, while a serialist approaches problem-solving step-wise, proceeding from the known to the unknown.

The cognitive style of individuals is usually stable over long periods of time, and is by many considered to be a personality trait. Even though the cognitive styles are not meant to fit a specific situation, they still have consequences for the resulting effectiveness. To take the last example, a holist can – if lucky – be efficient, but is rarely thorough. A serialist, on the other hand, is thorough and because of that usually not efficient. In addition to being cognitive styles, we can also recognise many of the same techniques or heuristics as solutions for specific situations.

Collective (Organisational) ETTO Rules

If we look to the organisation, it is possible to find collective counterparts to the individual ETTO rules. In the systemic view, which is the basis for this whole book, organisations are complex socio-technical systems that interact with and try to control a partly unpredictable environment.

- One rule is *negative reporting*, which means that only deviations or things that go wrong should be reported. In consequence of that, the absence of a report is interpreted as meaning that everything is well. The rule clearly improves efficiency, but may have consequences for safety.
- Another rule can be called the *prioritising dilemma* or the *visibility-effectiveness* problem. Many organisations realise that it is important for managers at various levels to be visible in the organisation, which means that they should spend time to find out what is going on and become known among the people they manage. On the other hand, managers are often under considerable pressure to be effective, to perform their administrative duties promptly even when deadlines are short. They are therefore required by their bosses to be both efficient in accomplishing their administrative duties, and thorough in the sense that they are good managers – i.e., highly visible. Managers will in practice often focus on efficiency (accomplishing their administrative duties) and trade-off thoroughness, being less visible in the organisation. If nothing untoward happens, he or she will be praised for the efficiency, but if something goes wrong, they will be blamed for their lack of thoroughness.
- *Report and be good.* Yet another example is in the relation between an organisation and a subcontractor or a supplier. Here the safety ethos prioritises openness and reporting of even minor mishaps. Subcontractors and suppliers thus often feel under pressure to meet the organisation's standards for openness and reporting. But at the same time they may have experienced, or believe they will experience, that they will be punished if they have too many things to report, while a competitor that reports less may be rewarded. In ETTO terms it is thorough to report everything and efficient to

report enough to sound credible but not so much that one loses the contract.

- *Reduce unnecessary costs.* While this may sound plausible enough at first, the problem lies with the definition of 'unnecessary.' The rule is used to improve efficiency, at the cost of thoroughness.
- *Double-bind* describes a situation where a person receives different and contradictory messages. A common example is the difference between the explicit policy that 'safety is the most important thing for us,' and the implicit policy that production takes precedence when conflicts arise. The double-bind that results from this is used to improve efficiency, at the cost of thoroughness. Another example is the visibility-effectiveness problem described above.

The ETTO rules described above offer a convenient way of classifying typical behaviours from either a work, an individual, or a collective perspective. They should not be construed as explanations in the sense that they represent underlying individual or organisational 'mechanisms.' Observable behaviour can be described 'as if' a person followed an ETTO rule, but it would be a mistake to assume that the ETTO rules constitute causes of behaviour. (In addition, to do so would itself be a case of ETTOing.)

Sources for Chapter 2

Post-decisional consolidation, also known as Differentiation and Consolidation theory, studies how people after a decision sometimes may re-evaluate the alternatives and restructure the problem, to maintain consistency between past and present. A good reference is the chapters collected in R. Raynard, W. R. Crozier and O. Svenson (eds) (1997), *Decision Making: Cognitive Models and Explanations* (London: Routledge). For an excellent description of the history of decision theory and rationality, see P. L. Bernstein (1996), *Against the Gods: The Remarkable History of Risk* (New York: John Wiley & Sons, Inc).

The classical treatment of the temporal side of actions is found in J. F. Allen (1984), 'Towards a general theory of action and time,' *Artificial Intelligence*, 23, 123–154, which provides a systematic account of the possible relations. Time has, however, generally been treated in a step-motherly fashion both in decision-making and in the study of human action because models and theories have focused on information–processing as an internal activity.

The study of foraging behaviour is part of ecology. The findings reported in this chapter are from N. B. Metcalfe and R. W. Furness (1984), 'Changing priorities: The effect of pre-migratory fattening on the trade-off between foraging and vigilance,' *Behavioral Ecology and Sociobiology*, 15, 203–206.

The 'Elimination By Aspects' (EBA) decision principle is a good representative of the field known as description decision theory, which was developed as an alternative to classical (normative) decision theory. Two of the leaders in this development were Amos Tversky and Daniel Kahneman. A full description of the EBA can be found in A. Tversky (1972), 'Elimination by aspects: A theory of choice,' *Psychological Review*, 79(4), 281-299. Similarly, prospect theory is described in D. Kahneman and A. Tversky, (1979), 'Prospect theory: An analysis of decision under risk,' *Econometrica*, 47(2), 263–291. The ground-breaking research by the same authors on the heuristics people rely on in judgement was first described in A. Tversky and D. Kahneman, (1974), 'Judgment under uncertainty: Heuristics and biases,' *Science*, 185, 1124–1131, and later became part of a collection of related papers in a book by the same name (Cambridge University Press, 1982).

The term 'double-bind' describes a situation in which someone, an individual or group, receives two or more conflicting messages such that one message effectively negates the other. It follows that the response can be classified as wrong, whatever it is. The term was used by the anthropologist Gregory Bateson (1904–1980), and described in his 1972 book *Steps to an Ecology of Mind* (University of Chicago Press).

Chapter 3: Explaining Human Irrationality

As already noted, it is not exactly news that humans habitually make a trade-off between efficiency and thoroughness. The social and behavioural sciences offer several different descriptions of this phenomenon, both on the level of the individual and on the level of the organisation – or rather for humans as psychological 'machines' and for humans in the context of an organisation or social environment. The following examples provide a chronological illustration of how widespread the recognition of the ETTO principle has been, without pretending to be exhaustive.

Satisficing

In 1955, Herbert Simon, a political scientist who in 1978 received the Noble Prize in economics, introduced *satisficing*, the term being a combination of *satisfy* and *suffice*, as a theory of human behaviour to explain the discrepancy between rational and real behaviour. Relative to economic theory, *satisficing* enables a decision-maker to achieve some minimum level of a particular variable, but not necessarily to maximise it. (The phenomenon is also known as *bounded rationality*.)

Many studies had unequivocally demonstrated that people did not conform to the requirements of rational decision-making (see Chapter 2), and for behavioural and organisational scientists this discrepancy cried out for an explanation. (Most scientists dislike something they cannot explain, cf. the Nietzsche quotation in Chapter 1.) Simon's ambition was to replace the rationality of *homo economicus* with a kind of rationality that corresponded to state-of-the-art psychological knowledge. At the time, in the mid-1950s, that was the beginning of the view of the human as an information–processing system. The ambition was therefore to prove that human decision-making was rational in terms of people's information–processing characteristics and the nature of the environment, even if it was not rational in terms of economic decision theory.

Without going into every detail of the theory, its basic premise was that rationality imposed a high cognitive load on the human mind, and that there was an innate tendency to reduce load, a kind of built-in regulatory mechanism or defence. According to Simon, the reason for *satisficing* was that humans did not have the cognitive resources to be able to maximise, in the sense that the processing or calculation required – the thinking and reasoning – would exceed their cognitive capacity. Instead of trying to find the best alternative, a decision-maker would stop as soon as an acceptable or satisfactory one was found. The theory was thus based on the common assumption of limited mental or cognitive capability.

Satisficing versus Sacrificing

Satisficing describes a trade-off between efficiency and thoroughness, but proposes that this trade-off is an inherent characteristic of the human mind. (This, of course, rests on the assumption that human cognitive capacity is limited *per se*.) It is, however, possible to interpret the very same behaviour in a different way, namely in terms of *sacrificing*. Sacrificing does not explain the lack of rationality in terms of the limited information–processing capacity of humans, but rather in terms of the intractability of the working environment. (A system, or an environment, is intractable if the principles of functioning are partly unknown, if descriptions are elaborate and detailed, and if it changes before a description can be completed. Tractability and intractability are described in more detail towards the end if this chapter.) In this view, too much time spent on thinking (on being rational) takes time away from doing – time that furthermore is of uncertain length because it is not known when it becomes necessary to do something else. In order to remain in control, humans – and most living organisms with them – therefore seem to prefer efficiency to thoroughness. Efficiency means getting something done in time or with a little time to spare, even if this means being a little less precise. Thoroughness means being as precise as possible, even if this means running the risk of being short of time or unable to respond when something unexpected happens. In other words, to the extent that the environment is intractable and to the extent that unexpected events may happen, it pays to make a sacrifice, to trade off thoroughness for efficiency. If, on the other hand, the environment is tractable, and future events therefore predictable, then there is little to be gained from sacrificing. In this case it is known when something must be ready, and sufficient thoroughness can therefore be achieved – at least in principle.

Whereas *satisficing* is explained as a consequence of limited cognitive capacity, *sacrificing* is explained as a consequence of the intractability of the work environment. The tractability is reduced for three main reasons (cf. Figure 3.1). First, systems grow larger so there are more parts and more details to account for. Second, the rate of change has increased, both because processes have become faster (*vide* the computer revolution and Moore's law), and because they are more tightly coupled so that the consequences of a change in one place will propagate faster and spread to a larger part of the system. Third, performance demands have increased, partly in response to the promises of the more powerful technology.

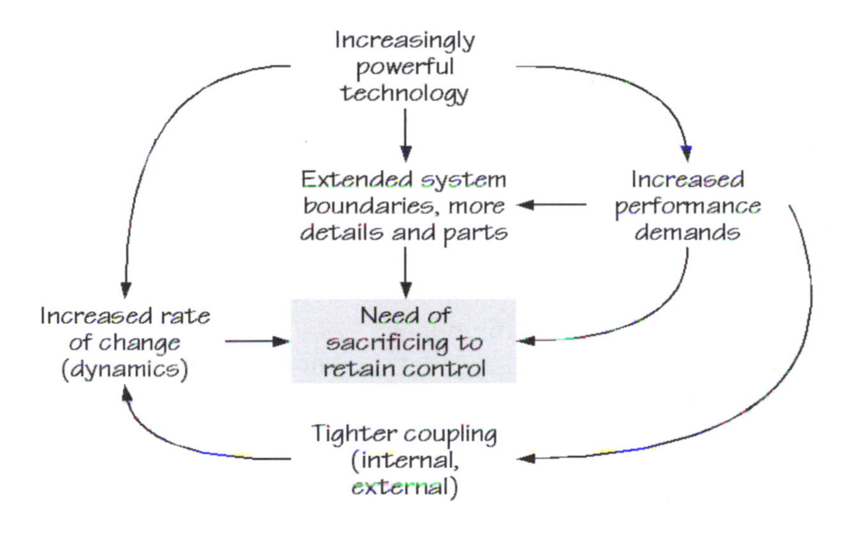

Figure 3.1: Forces affecting system growth

These developments altogether lead to a greater pressure towards efficiency, hence less possibility of being thorough. The increased pressure for efficiency coupled with the faster rate of change and the faster pace of the processes we deal with, means that we need to sacrifice something in order to overcome the need for both efficiency and thoroughness. And as usual we sacrifice by cutting down on thoroughness, at least until we reach the level where it no longer feels safe.

While *satisficing* and *sacrificing* initially appeared to describe the same phenomenon, on further reflection they do not. *Satisficing* ascribes the human inability to be rational to limited processing capacity. *Sacrificing* describes how humans cope with the complexity of their environment, not least the temporal complexity, by deliberately (and later on habitually) making a trade-off, usually between thoroughness and efficiency. That is in itself a very rational thing to do, perhaps not in the normative sense but in the sense of being able to maintain control. There is, indeed, little merit in being normatively rational if it means that the existence of the system is put in danger.

Muddling Through

In 1959 Charles Lindblom, who was professor of political science and economics at Yale University, published a paper entitled *The Science of 'Muddling Through.'* (A dictionary definition of 'muddling through' is to 'to push on to a favourable outcome in a disorganized way.') Decision theories usually prescribe a rational approach to decision-making that involves the identification of alternatives, the comparison of alternatives, and finally the selection of the optimum alternative. Even descriptive decision theories such as *satisficing*, *Elimination By Aspects*, and *prospecting* try to do so. Lindblom called this approach the *rational–comprehensive* or root method. In practice, however, people more often make decisions by defining the principal objectives, outlining the few alternatives that occur to them, and finally selecting one that is a reasonable compromise between means and values. Lindblom called this the *successive limited comparisons* or branch method. The two methods are summarised in Table 3.1. Lindblom argued, by means of examples of decision-making by administrators, that since the successive limited comparisons method often was a common method of policy formulation, it should not be seen as a failure of the rational method but rather as a method or a system on its own.

In relation to the ETTO principle, the rational–comprehensive method represents thoroughness while the successive limited comparisons method represents efficiency. The rational–comprehensive method requires considerable time and resources if it is to be done right, and probably exceeds what most people can do reliably, even with the help of so-called intelligent decision support. The successive comparisons method sets more realistic requirements to the policy maker by breaking the problem or decision into a number of steps. In fact, as the name implies, the decisions are taken successively, which means that the demands from each step are smaller.

Table 3.1: Comparison of two decision-making methods

The rational–comprehensive (root) method	**The successive limited comparisons (branch) method**
Clarification of values or objectives are distinct from and usually prerequisite to empirical analysis of alternative policies.	Selection of value goals and empirical analysis of the needed actions closely intertwined.
Policy-formulation is approached through means-end analysis.	Means and ends are not distinct, and means-end analysis is often limited.
The test of a 'good' policy is that it can be shown to be the most appropriate means to desired ends.	The test of a 'good' policy is typically that various analysts find themselves directly agreeing on a policy.
Analysis is comprehensive; every important relevant factor is taken into account.	Analysis is drastically limited and important possible outcomes, alternative policies, and affected values are neglected.
Theory is often heavily relied upon.	Successive comparisons greatly reduce reliance on theory.

The 'muddling through' proposes that people usually achieve their goals (solving a problem, making a decision, carrying out a plan) by actions which are more like stumbling than the deliberate execution of a well-defined procedure. Indeed, actual human performance usually differs markedly from the elegant descriptions that are implied by most theories of cognition. In practical actions humans are up against a number of pervasive features of daily life such as limited attention, the inherent variability of action, information input overload, bounded memory capacity, ambiguity in communication, uncertain resources, unpredictable interruptions, etc. It is, perhaps, because of this that performance seems to be a constant compromise between goals, resources, and demands, thereby giving rise to the 'muddling through.'

Recognition Primed Decision-Making

While *satisficing* and *muddling through* in their separate ways purported to describe what decision-makers actually did, neither provided a psychologically satisfactory explanation. In the 1970s and 1980s, however, the focus on decision-making as a process in its own right subsided, and different aspects of human performance came into focus. Another change was that while decision-making in the 1950s mainly was seen as something done by managers and decision-makers of various kinds, i.e., typically people working at the so-called blunt end of the systems, the technology-induced changes to the nature of work meant that others had to face decisions as well. In 1989 a workshop was held in Dayton, OH, which started the tradition that is now known as *Naturalistic Decision-Making* (NDM). In distinction to formal decision-making, NDM is the study of decision-making, sense making, and planning in natural settings, for instance by fire fighters, surgeons, etc. While these people are not professional decision-makers in the way that people at the blunt end supposedly are, their decisions are nevertheless a critical part of their activities. The purpose of NDM is to understand how people make rapid decisions in situations characterised by time pressure, ambiguous information, ill-defined goals, and changing priorities – in other words, in real-life conditions – since this is something that clearly exceeds the scope of rational decision theory. The fact that the decisions are rapid makes even Lindblom's successive comparisons method inappropriate because there rarely is time or opportunity to make more than the first decision.

The proposal of NDM is that people rely on *recognition-primed decisions* (RPD). According to this, the decision-maker tries to match the characteristics of the current situation with a generic situation and use that to come up with a possible course of action. This is then compared to the constraints imposed by the situation, to determine if it is good enough. If that is not the case, it is rejected and a new possible course of action is generated. In relation to ETTO, RPD is clearly a trade-off of thoroughness for efficiency and in this case a rather deliberate one. (Not only does RPD as a whole represent ETTO, but the functions or steps of RPD do as well.) The basis for the decision is the recognition that an action makes sense in the situation together with the quick judgement that the action may produce an acceptable outcome. The fact that the situation changes rapidly, as in fire fighting, means that there is little if any advantage in making a more deliberate comparison of alternatives. The dynamics of such situations makes them inherently intractable, hence it renders any kind of rational or contemplative solution inappropriate. Under such conditions, doing is usually better than thinking.

The Use of Schemata

The idea that people form or create some kind of description of a situation and use this as a reference for what they do, is well-known in psychology where it goes by the name of *schema*. A schema is usually defined as a mental representation of some aspect(s) of the world, i.e., of some important characteristics of a situation. The use of the schema means that a situation or a condition can be 'understood' by simply checking it against the characteristics described by the schema, rather than by looking at all the details and try to make sense of them as a whole (cf. also the description of *Information Input Overload* below). The earliest known usage of the term in psychology is by the Swiss philosopher and developmental theorist Jean Piaget, ca. 1926, and by the British psychologist Sir Frederic Bartlett, ca. 1932. Far earlier than that a similar idea was used in philosophy, for instance Plato, who developed the doctrine of ideal types, or Immanuel Kant, who developed the notion of a schema or a mental pattern.

The idea of a schema was in the 1970s and 1980s a central part of the theories of humans as information–processing systems, of artificial intelligence (where it was called a *frame*), and of the psychology of perception. A schema is an efficient way to understand a situation since it obviates the need for effortful thought. Instead it is sufficient to recognise the situation, as in RPD. In decision-making where time is short and the requirement to do something is strong, it is clearly efficient to see if the current situation is similar to one that is already known ('*it looks like Y, so it probably is Y*'), and then to use the same response or the same principles as a basis for action ('*it normally works*'). As long as the situations really are similar, the likelihood of an adverse outcome is small. The largest risk probably comes from having too simple a pattern or a schema as a basis. In the extreme, a schema may be reduced to a binary classification, such as 'one of us' or 'one of them,' with all the drawbacks this carries with it.

The greatest problem with a schema is that it can lead to what is technically called a negative transfer. This means that a response (or a decision to act in a certain way), that has been found appropriate for one type of situation, is transferred to or used in situations where it is inappropriate. Transfer, or positive transfer, is not only efficient but actually essential for our ability to cope with the complexity of the situations we encounter: by making use of what we have learned, we eliminate the need to assess each and every situation from scratch. The negative transfer, i.e., the use of a learned response when it is not appropriate, happens because two situations superficially seem to be similar or even identical, even though they are not. A relatively innocent example is when you rent a car where the placement of the controls for the indicators and the windshield wipers is the opposite of what you are used to. Or the difference between calculator keypads, which usually start with '7-8-9' in the upper row, and telephones, which start with '1-2-3' instead. Such inconsistencies between otherwise similar pieces of equipment may in the worst case give rise to accidents, as the following example demonstrates.

This accident involved an unmanned aircraft (UA), which on 25 April 2006 crashed following a loss of engine power during a mission patrolling the southern US border. The report by the US National Transportation Safety Board (NTSB) explains it as follows:

> The flight was being flown from a ground control station, which contained two nearly identical control consoles: PPO-1 and PPO-2. Normally, a certified pilot controls the UA from PPO-1, and the camera payload operator ... controls the camera, which is mounted on the UA, from PPO-2. Although the aircraft control levers ... on PPO-1 and PPO-2 appear identical, they may have different functions depending on which console controls the UA. When PPO-1 controls the UA, movement [of] the condition lever to the forward position opens the fuel valve to the engine; movement to the middle position closes the fuel valve to the engine, which shuts down the engine; and movement to the aft position causes the propeller to feather. ... [T]he condition lever at the PPO-2 console controls the camera's iris setting. Moving the lever forward increases the iris opening, moving the lever to the middle position locks the camera's iris setting, and moving the lever aft decreases the opening. Typically, the lever is set in the middle position. ... Console lockup checklist procedures indicate that, before switching operational control between the two consoles, the pilot must match the control positions on PPO-2 to those on PPO-1 by moving the PPO-2 condition lever from the middle position to the forward position, which keeps the engine operating. ... [D]uring the flight, PPO-1 locked up, so (the pilot) switched control of the UA to PPO-2. In doing so, he did not use the checklist and failed to

> match the position of the controls on PPO-2 to how they were set on PPO-1. This resulted in the condition lever being in the fuel cutoff position when the switch to PPO-2 was made, and the fuel supply to the engine was shut off.

It is hardly surprising that the pilot traded off thoroughness for efficiency when he had to switch from one console to the other in the middle of a flight. But in combination with the fact that the condition lever on the two consoles controlled different parts of the UA (fuel valve and iris opening, respectively), his schema for how to control the UA was inappropriate for the new situation. It is thus a characteristic example of negative transfer. Some people would argue that this was really a design failure, possibly due to some unknown ETTO during the construction of the consoles. The NTSB, however, concluded that the probable cause of the accident was the pilot's failure to use checklist procedures when switching from one console to another. The pilot should, of course, in hindsight have been thorough rather than efficient. But so should probably also a number of other people at the blunt end of this system, who instead were saved by the NTSB's 'stop rule.'

Speed–Accuracy Trade-Off

There is generally a relation between how fast something can be done, i.e., the speed, and how well or how precisely it can be done, i.e., the accuracy. This is known in psychology as the speed–accuracy trade-off. The two extremes are that a user either can be very fast with a large number of errors or very slow with very few errors. When asked to perform a task as well as possible, people will apply various strategies that may optimise speed, optimise accuracy, or combine the two. For this reason, comparing the performance of two users cannot be done on the basis of speed or accuracy alone, but both values need to be known. (Experimental psychologists use various types of controlled tasks such as pointing to items, pressing buttons in response to a signal or following a movement, recognising confusing or unclear signs, etc. The parlour game version is the famous 'tricky fingers' knife trick.)

More technically, the speed–accuracy trade-off can be illustrated as in Figure 3.2 below. Here the X-axis shows the time used to do something, for instance the time it takes to answer which signal was presented, and the Y-axis shows the accuracy. It stands to reason that in cases where the signal was not clear, for instance a text that is hard to read, then the more time you can spend on it, the more likely it is that the response will be correct. The trade-off happens because there is a penalty for taking too much time as well as a penalty for answering wrongly. For a concrete example, think of a radar operator trying to identify an approaching object, which could be friend or foe. An incorrect identification might be fatal, but so might waiting too long in order to be sure.

Speed–accuracy was of considerable concern for the Scientific Management movement and later for the Methods-Time Measurement (MTM) system that was used to quantify the amount of time required to perform specific tasks. It obviously is desirable (for the organisation) to increase the speed of work, as long as it does not affect accuracy. There is usually a level above which an increase in speed will have a negative effect on accuracy. Conversely, there is a level below which a reduction in speed will not lead to an improvement in accuracy. The trick is to find the right balance between these limits.

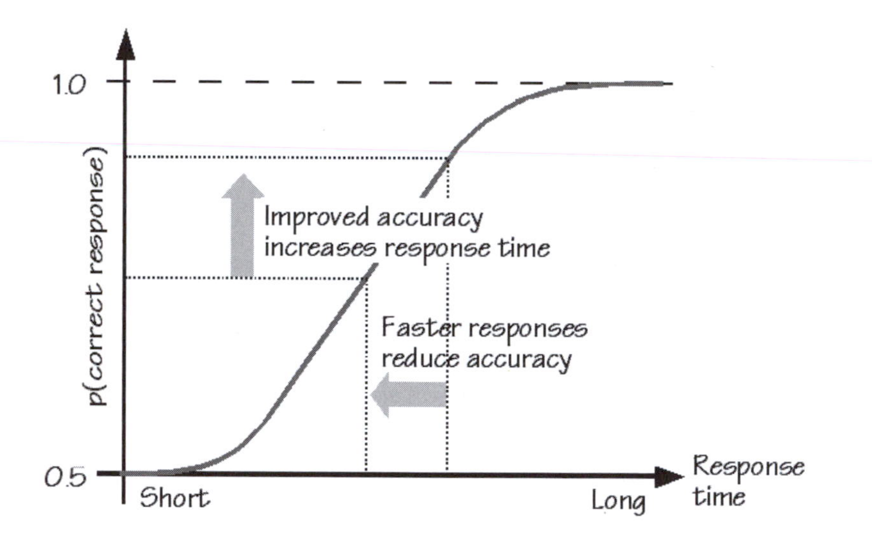

Figure 3.2: The speed–accuracy trade-off

The speed–accuracy trade-off was given a precise mathematical formulation by Paul Fitts in 1954, and thence became known as Fitts' law. The law predicts the time required for a rapid, aimed movement of, say, a lever or a computer mouse, from a starting position to a final target area. The time is a function of the distance to the target and the size of the target. Fitts' law is an effective method of modelling rapid, aimed movements, where, e.g., a hand starts at rest at a specific start position and moves to rest within a target area. It can be used to assist in user interface design, predict the performance of operators using complex systems, and predict movement time for assembly line work.

The speed–accuracy trade-off can also be used to characterise phenomena on a completely different level, such as organisational changes. In these cases it is not unusual that the pressure to plan and implement a change – often with the primary purpose to cut costs – forces decisions to be made on an insufficient basis. Examples from major industrial domains around the world are not hard to find, e.g., the way the European Aeronautic Defence and Space Company (EADS) responded to the delays in the production of the A380 aircraft, or the way the US automobile industry responded to the change in consumer needs during the economic crisis in 2008. The speed–accuracy trade-off is sometimes also invoked deliberately, as when an 'adversary,' e.g., a labour union, is given very little time to respond to a complex proposal on the expectation that accuracy or thoroughness will be traded off for speed or efficiency in order to respond before the stipulated deadline.

Time–Reliability Correlation

Yet another attempt to formalise or systematise the relation between speed and accuracy is provided by the time–reliability correlation (TRC). The TRC considers the quality of performance in terms of whether a person responds correctly in a demanding situation. It is quite reasonable to assume that it may be difficult immediately to respond correctly when faced with a demanding situation or event. It also makes sense to assume that the likelihood of responding correctly increases as a function of time after the beginning of the event, although not necessarily in a linear fashion. Time provides an opportunity to build an understanding of the situation as well as to gather additional data, test hypotheses, etc.

The prototypical situation describes an operator in a nuclear power plant control room faced with an unexpected disturbance – or in principle any other person facing a disturbance at work. Since a correct response is essential for safety, there has been considerable concern about how likely this is, and the TRC was an early attempt to quantify human reliability in a simple manner. For some reason the TRC has traditionally expressed the probability of not performing an action as a decreasing function of the time available for that action, and was typically shown by means of a logarithmic plot with time on the X-axis and the non-response probability on the Y-axis (Figure 3.3).

In this way the probability of responding (correctly) increases as time goes on, which corresponds to the common-sense notion that sooner or later something will be done. The available time, however, refers to the time that has elapsed since the beginning of the event, rather than the subjectively available time.

In relation to efficiency and thoroughness, the TRC expresses the reasonable assumption that greater thoroughness, assumed to be a consequence of more time, will lead to a better response. In this case the emphasis is therefore on thoroughness rather than efficiency. The reason is that in a safety critical situation, accuracy can be more important than speed.

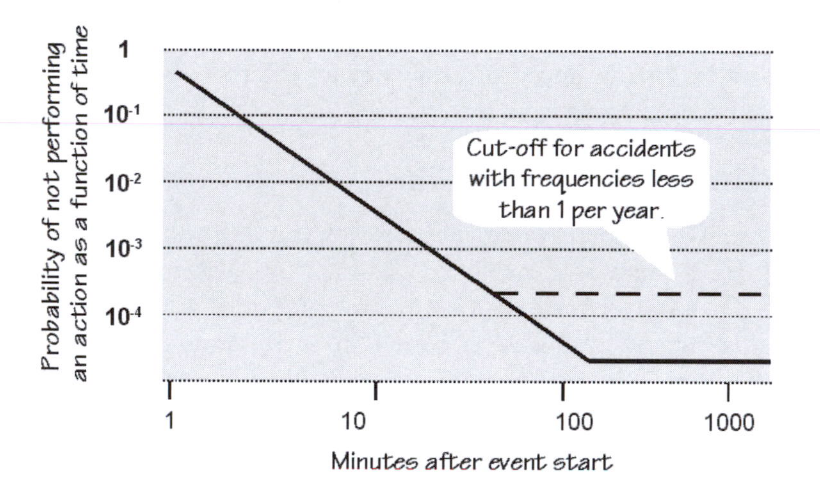

Figure 3.3: The time–reliability correlation (TRC)

The TRC can be seen as a formal acknowledgement of the relation between efficiency and thoroughness (or rather speed and accuracy), in a way that can be used for safety calculations and work design, and also finally as a recognition that in some situations efficiency or speed is not the most important thing – at least not for a while.

Information Input Overload

Thoroughness requires that all, or as much as possible, of the available information is used (processed or taken into account). This demand is reasonable, as long as the amount of information is limited. But in our day and age that is not the case. We are all constantly inundated with data (information), and frequently find ourselves in a situation that best can be characterised as *information input overload*.

Information input overload, or just information overload, means that there is more information available than can be made use of in the time available. This can happen either (1) because there is an increase in the rate of incoming data, for instance during the early parts of a disturbance in a process plant, or (2) because the capacity available to take care of or process the data is reduced – temporarily or permanently. Modern technology makes it possible to collect prodigious amounts of data, as e.g., in networked enabled defence, stock exchanges, or satellite measurements of the weather, but the problem is what to do about it. The dilemma is that it may be necessary to do something before all the data have been considered. In other words, there is a possible conflict between efficiency and thoroughness.

While too much information is a situation that most people in the industrialised world are familiar with, at work or at leisure, it is by no means a new phenomenon. In fact, the term *information input overload* was proposed already in 1960 by the biologist James Grier Miller, well before the IT revolution had become ubiquitous. In a survey of the then available literature Miller proposed a small set of typical ways of reacting to information input overload, i.e., to the situations where a person had to deal with more information than he could handle. The names of the typical reactions are practically self-explanatory (Table 3.2), and still highly relevant today.

The reactions to information input overload can be seen as a trade-off between efficiency – getting the task done – and thoroughness – getting the essential information for the task. The problem has grown a lot worse since 1960, and technological developments have, ironically, both increased the problem but also helped with some of the solutions. The problem can be expressed in different ways. For example, it could be stated as the person's strategy for maintaining an appropriate representation of the situation so that it always corresponds reasonably well to the state of the world. It might also be described as the ways in which people cope with the complexity of the world by reducing the complexity in various ways to match the available needs and capabilities, thereby maintaining a functioning homoeostasis.

Table 3.2: Characteristic reactions to information input overload

Solution	Criterion	Description
Omission	Complete the task without further disturbances.	Temporary non-processing of information, regardless of type or category.
Reduced precision	Work faster but without missing essential information.	Trading accuracy for speed and/or time. This will lead to a more shallow use of the input information.
Queuing	Do not miss any information (only efficient for temporary conditions).	Delaying response during high load, hoping for a lull later on. Save incoming data for later.
Filtering	When time/capacity restrictions are severe.	Neglecting to process certain categories, prioritising data types.
Cutting categories	When time/capacity restrictions are severe.	Reducing number of categories (e.g., friend vs. foe).
Decentralization	Do not miss any information.	Distribute processing if possible, employ additional resources.
Escape	Survival, self-preservation.	Abandon the task, flee the field.

The very notion of these reactions implies a coupling between what a person knows or assumes to be the case and the actual situation. Essential information is not an objectively defined category, but depends on the currently salient features of the person's representation of the situation. Or rather, the determination of what is the essential information and what is not is based on the person's current understanding of the situation and their current goals. There is therefore a rather tight coupling between the information that is potentially available, the information that the person looks for, and the way in which this information is treated.

This can be illustrated by looking at the ETTO principle as a decision strategy. This is often done by combining different types of outcome (success, failure) with different strategies (being efficient, being thorough) and showing the outcome as a 2×2 matrix, cf. Table 3.3.

Given that positive outcomes normally outnumber adverse outcomes by several orders of magnitude, people who change from being thorough to being efficient will be reinforced because they move from the *misgivings* cell to the *relief* cell. Should things actually turn out wrong, people may *regret* being efficient instead of being *vindicated* for being thorough, but the difference between the two is not large enough to matter in practice.

Table 3.3: The ETTO principle as a decision strategy

	Situation as assumed (positive outcome)	Situation not as assumed (adverse outcome)
Trade-off thoroughness for effectiveness	Relief: Success, even though efforts were saved.	Regrets: Failure, should have been thorough.
Trade-off effectiveness for thoroughness	Vindication: Success, wise use of effort.	Misgivings: Failure, and effort spent in vain.

ETTO as a Principle of Rationality

As this chapter has shown, the recognition that people are not rational decision-makers is quite old and the very attempts to formulate principles of logic or rationality may be seen as an implicit recognition of that. Philosophers and scientists have through the ages tried to formulate principles that, if followed, would ensure rationality. (The meaning of rationality is, of course, itself a relative and vaguely defined concept or criterion. Here it is used in the meaning of 'sound,' i.e., 'thorough and complete.') Some of the more important ones have been described above.

It may be asked whether the ETTO principle should be considered as yet another proposal to save rationality, to make it scientifically palatable. The answer to that is a clear *no*. The situation is, indeed, quite different. The ETTO principle names a phenomenon or a strong characteristic of individual – and collective – performance, namely that people in dynamically changing, hence unstable and partly unpredictable situations, know that it is more important to do something before time is up, however imperfect it may be, than to find the perfect response when it is too late. The ETTO principle is a good description in the sense that it enables us better to understand what other people do or may do, hence enable us to be sufficiently efficient in what we do ourselves. It does not explain why people make their trade-offs in any kind of deep, scientific or philosophical sense. It names a phenomenon, and by doing so becomes an instance of what it names. So even if ETTO is not another principle for rationality, there is a certain rationality in ETTOing, because it enables us to maintain control in a changing and uncertain world.

People (humans and organisations) do not ETTO as a deliberate choice – at least not normally. They ETTO because they have learned that this is an effective strategy – by imitation or by active encouragement. They continue to do so because they are rewarded or reinforced in using the heuristic. They are warned not to do it when they fail, depending on how the experience is interpreted (my fault, someone else's fault?). But since success is normal and failure is rare, it requires a deliberate effort not to ETTO, to go against the pressure of least effort.

Tractable and Intractable Systems

In order for a system to be controllable, it is necessary that what goes on 'inside' it is known and that a sufficiently clear description or specification of the system and its functions can be provided. The same requirements must be met in order for a system to be analysed, so that its risks can be assessed. That this must be so is obvious if we consider the opposite. If we do not have a clear description or specification of a system, and/or if we do not know what goes on 'inside' it, then it is clearly impossible to control it effectively as well as to make a risk assessment. We can capture these qualities by making a distinction between tractable and intractable systems, cf. Table 3.4. A system is tractable if the principles of functioning are known, if descriptions are simple and with few details and, most importantly, if it does not change while it is being described. An example could be an assembly line or a suburban railway. Conversely, a system is intractable if the principles of functioning are only partly known (or in extreme cases, completely unknown), if descriptions are elaborate with many details, and if systems change before descriptions can be completed. An example could be emergency management after a natural disaster or, *sans comparison*, financial markets.

It follows directly from the definition of tractability that an intractable system also is underspecified. The consequences are both that the predictability is limited and that it is impossible precisely to prescribe what should be done. Underspecification is, of course, only an issue for the human and organisational parts of the system. For the technical parts, in so far as they work by themselves, complete specification is a necessity for their functioning. An engine or a machine can only function if there is a description of how every component works and how they fit together. It is in this sense that the performance of a technical system *results* from the parts. Technological systems can function autonomously as long as their environment is completely specified and preferably constant, in the sense that there is no unexpected variability. But this need of a complete technical specification creates a dilemma for socio-technical systems. For such systems the environment cannot be specified completely and it is certainly not constant. In order for the technology to keep working, humans (and organisations) must function as a buffer both between subsystems and between the system and its environment, as something that absorbs excessive variability when there is too much of it and provides variability when there is too little. The problem can in some cases be solved by decoupling parts of the system, or by decomposing it. But for an increasing number of systems this solution is not possible.

Table 3.4: Tractable and intractable systems

	Tractable system	Intractable system
Number of details	Descriptions are simple with few details	Descriptions are elaborate with many details
Comprehensibility	Principles of functioning are known	Principles of functioning are partly unknown
Stability	System does not change while being described	System changes before description is completed
Relation to other systems	Independence	Interdependence
Controllability	High, easy to control	Low, difficult to control
Metaphor	Clockwork	Teamwork

Sources for Chapter 3

The thinking behind satisficing was first described by Herbert A. Simon in a paper from 1955, 'A Behavioral Model of Rational Choice,' *The Quarterly Journal of Economics*, 59(1), 99–118. The first use of the term satisficing came a year later in another paper by Simon, 'Rational choice and the structure of the environment,' *Psychological Review*, 63(2), 129–138. C. E. Lindblom's description of 'muddling through' came in 1959, in a paper entitled 'The science of 'muddling through,' *Public Administration Review*, 19, 79-88.

The first book about naturalistic decision-making is G. Klein, J. Orasanu, R. Calderwood and C. E. Zsambok, (1993) *Decision Making in Action: Models and Methods,* (Norwood, NJ: Ablex Publishing Co). The details of the recognition-primed decision model were described in a later volume. The full reference is G. Klein, (1997), 'The recognition-primed decision (RPD) model: Looking back, looking forward,' in C. E. Zsambok & G. Klein (eds), *Naturalistic Decision Making,* (Mahwah, NJ: Lawrence Erlbaum Associates). The series of naturalistic decision making conferences that began in Ohio in 1989 are still ongoing (pp. 285–292).

The use of schema, or schemata, in psychology goes back to Swiss philosopher and development theorist Jean Piaget (1896–1980) in 1926, while the use in philosophy goes back to Plato. The notion of a schema is today widely used in both psychology and Artificial Intelligence. Transfer of learning, including negative transfer, has been part of experimental psychology since the beginning of the 20th century. Both schema/schemata and transfer have been dealt with extensively in psychology, in both introductory and advanced texts. The same goes for the speed–accuracy trade-off. The accident with the unmanned aircraft is described in Report No. CHI06MA121 from the National Transportation Safety Board (http://www.ntsb.gov).

The Time–Reliability Correlation (TRC) was one of the many methods that were developed to solve the problem of Human Reliability Assessment (HRA) after the Three Mile Island accident in 1979. An early description is in a report from the US Nuclear Regulatory Commission: R. E. Hall, J. Fragola and J. Wreathall, (1982), *Post Event Human Decision Errors: Operator Action Tree/Time Reliability Correlation* (NUREG/CR-3010). A comprehensive overview of representative HRA methods can be found in B. Kirwan, (1994),

A Guide to Practical Human Reliability Assessment (London: Taylor & Francis).

The term information overload is often attributed to Alvin Toffler's book *Future Shock* from 1979. But the term was actually introduced almost two decades earlier by J. G. Miller, (1960), 'Information input overload and psychopathology,' *American Journal of Psychiatry*, 116, 695–704. The whole idea about information overload is based on the analogy with an information–processing system, where limitations can arise from either channel capacity or processing capacity. Miller's analysis provided a set of terms that is still applicable today, and the original paper is certainly worth reading.

Chapter 4: Efficiency-Thoroughness Trade-Off in Practice

The Choice of Efficiency

The first three chapters have introduced the ETTO principle and described some forms of this pervasive phenomenon. As these chapters have suggested, there is no shortage of real-life situations where the ETTO principle can be found. Indeed, because it is such a nearly universal characteristic of human performance, it may actually be hard to find an example on either an individual or collective level that does not exemplify it in one way or the other. While this chapter will present further examples from various domains, the main purpose of this book is not to convince anyone that the ETTO principle is real in a philosophical sense. It is rather to consider the consequences, in particular for safety, of this way of describing what humans and organisations do. It is only by having an adequate understanding of the nature of individual and collective performance that we can ever hope to be able to control it. And control is necessary to be safe – or for that matter, to be efficient.

Even though the ETTO principle describes a phenomenon that is ubiquitous, it does not mean that it always is automatic or unconscious – or even unintentional. Indeed, as several of the following examples will show, people may often deliberately choose efficiency – and more unusually, thoroughness. Whereas the ETTO principle is easiest to recognise in individual performance, the very same phenomenon may be found when many individuals work together, in what we call organisations.

An organisation is, of course, composed of individuals and whatever the organisation does is done by the individuals in it. But even though it will always be an individual or a team of individuals that in the end makes the decisions, it is in some cases reasonable to talk about or refer to the organisation's performance, since the performance in question is a pervasive trait of the organisation rather than of any single individual. This may be because people are anonymous in the organisation, or because individual performance is dominated by organisational culture and social norms. An organisation may have its own agenda and prioritise efficiency over thoroughness, as can be seen in many cases – too many to mention here. It may even happen in the very large organisations that we call nations, as when the US administrations in 2001 claimed that following the Kyoto protocol would strangle US industry, hence in the face of global warming favoured efficiency for thoroughness. In many cases thoroughness is indeed discarded not out of ignorance but as a deliberate choice, i.e., it is decided to put the priority on effectiveness.

The purpose here is not to make a value judgement of any specific efficiency-thoroughness trade-off or sacrifice made (although it sometimes is very tempting to do so). Making such trade-offs is not only normal for humans and organisations, it is actually necessary. The best illustration of that is when people stop making them, as when they work strictly according to the written procedures and follow safety or other regulations to the letter. 'Work-to-rule' invariably leads to a slowdown and a loss of efficiency and is therefore often used as a minimal form of a labour strike. (Wikipedia defines work-to-rule as: 'an industrial action in which employees do no more than the minimum required by the rules of a workplace, and follow safety or other regulations to the letter in order to cause a slowdown rather than to serve their purpose.') In extreme cases, 'work-to-rule' is referred to as malicious compliance. The 'work-to-rule' phenomenon is a good illustration of why it is impossible in practice to maintain a high level of thoroughness, since this will bring things to a halt. It follows that rules and regulations on the whole are inappropriate for the work situation if strict compliance is required. This state of affairs is usually accepted in normal situations, but when something goes wrong it is frowned upon. The evaluation or assessment of performance is thus ambivalent: under normal conditions one set of criteria is used and under abnormal conditions another. It is OK to disobey the rules when things go right, but not when things go wrong.

The examples that follow show how the ETTO principle can be found in any situation and any kind of activity. Some describe accidents, others describe 'normal' work situations.

Ice Problems at Gardemoen

In the early morning of 27 December 1991, SAS Flight 751 took off from the Stockholm-Arlanda Airport in Sweden to fly to Copenhagen and then on to Warsaw. After 25 seconds of flight, the flight crew noticed noise and vibrations from one engine and responded by throttling down. Problems with the other engine began 39 seconds later, with both engines failing at 76 and 78 seconds into flight, at 3,000 feet of altitude. The pilot kept the aircraft gliding while attempting to restart the engines. As this failed, he chose to make an emergency landing. The plane hit some trees before touching down, lost a large part of the right wing and broke into three parts before coming to a stop on a field. Miraculously, no one was killed although two people were seriously injured. The engine failures leading to the crash were later found to have been caused by ice from the wings that had entered both rear-mounted engines. The ice had formed during the night before when the temperature dropped below freezing point, but had not been noticed by the maintenance crew.

After this accident, known as the Gottröra accident, after the place where the emergency landing took place, deicing procedures were re-evaluated and modified. In other words, the thoroughness of this particular procedure was emphasised. As a part of that, pilots were reminded of the importance of a full warm up of the engines before take-off.

Such thoroughness makes sense, both because the risk of a severe accident is substantial and because ice sucked into the engines may damage them, and hence necessitate expensive repairs. On the other hand, it takes time to follow these procedures. An immediate concern – to the airline and to the pilots – is that the pre-flight preparations may lead to delays or even to a flight missing its slot time. The pilots are therefore faced with a rather clear efficiency-thoroughness dilemma. If they choose thoroughness, they may run the risk of introducing delays, which no one, including passengers, like. If they, on the other hand, choose efficiency, they run the risk of damaging the engines or even of having a major accident.

Seven years after the Gottröra accident, on 14 December 1998, the engines of six SAS planes at Oslo Gardemoen Airport were ruined after icing. On the same day 15 more planes, from other airlines, suffered similar damage. Five years later, in February 2003, six SAS-owned planes suffered engine damage at the same airport and for the same reason. According to the Norwegian newspaper, *Aftenposten* (12 April 2004), a confidential internal report from the airline found that the pilots had neglected to warm up the engines before take-off. The airline admitted that it had at times not focused enough on the problem of icing in engines but rejected any implication that this had resulted in serious safety problems.

From an ETTO perspective, this case illustrates the rules of 'we must be ready in time' and 'it is good enough for now.' The first affects the second, in the sense that the need to reach the runway before the slot time has expired sets the limit for how long the warm-up of the engines can take. This should not come as a surprise to anyone, neither at the sharp nor at the blunt end.

The Leaking Barrel

A container with radioactive material for industrial use (366 terabequerel Iridium-192) was transported by land and air from Studsvik Holding AB in Nyköping, Sweden, to the Source Production and Equipment Company (SPEC), a leading manufacturer of industrial gamma radiography equipment in New Orleans (LA), USA. The container left Studsvik on 27 December 2001 and arrived in New Orleans on 2 January 2002.

The container was shaped like a barrel lying on its side and its extended measurements were 0.43×0.54 m. The travel of the container is shown in Figure 4.1. The container was sent by truck from Studsvik to the Stockholm-Arlanda airport from where it was flown to Memphis, TN, via Paris, France. The shipment was reloaded twice, once in Norrköping on the way to Stockholm-Arlanda airport, and once in Paris at Charles de Gaulle airport. After arriving in Memphis it was loaded on a truck and driven to New Orleans, where it was loaded onto a smaller truck for a short 10-minute final drive to the SPEC premises.

Directly upon arrival at SPEC, measurements were carried out to determine the dose rate. According to information from the US Nuclear Regulatory Commission, the measured dose rate from the side of the barrel (perpendicular to the barrel axis) was 1 Roentgen per hour at 15 feet. The reading from the lid of the barrel was 300–400 milliroentgen per hour at 75 feet. Readings at the bottom (opposite the lid) were minimal. (For comparison, acute radiation damage occurs at a dose of around 500 millisievert, with the normal annual dose in Sweden being about 4 millisievert. A dose of 400 milliroentgen corresponds to 47.7 millisievert.)

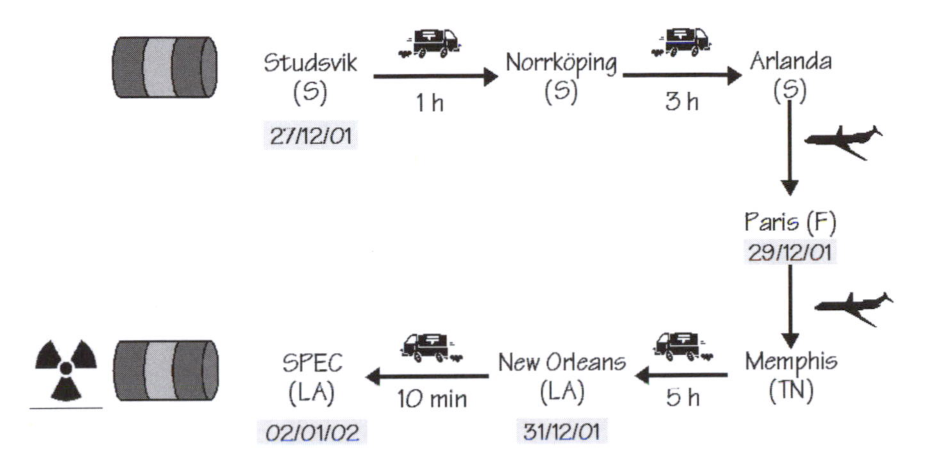

Figure 4.1: The container's journey

An explanation of how this happened includes examples of ETTO rules at many levels. A number of trade-offs were made in the filling and locking of the barrel at Studsvik, in the transportation of the container to the airport, and during the transportation from Stockholm to Memphis. The discussion here will, however, be restricted to the last part of the container's voyage, from the pick-up point in New Orleans to SPEC.

After the container arrived from Memphis it was picked up in Louisiana by a driver from the customer (SPEC). Before he loaded the container onto their truck, he tried to measure the local radiation level. According to the NRC Abnormal Event Report, the instrument showed the needle in a 'stuck' position. Although the report does not say whether it was at the minimum or maximum of the scale, it was probably the latter. Faced with this situation, the SPEC driver could in principle do either of two things. The first, which would emphasise thoroughness, would be to drive back to the company site, get a new instrument – and possibly some assistance – then drive back to the airport and repeat the measurements. Since the distance was only 10 minutes, it would in the worst case have delayed the delivery by, say, 30–60 minutes.

The second alternative was to reason as follows (note that this is conjecture, but hopefully not unrealistic): (1) It has never happened before that the needle was 'stuck' (ETTO rule: 'It looks like instrument failure, so probably it is instrument failure'); (2) the distance between the pick-up point and the company is too short for any harm to be done (ETTO rule: 'It is not really important'); and (3) it is more likely to be a failure of the instrument than a problem with the shipment (ETTO rule: 'It is normally OK').

Predictably, the driver chose the second line of action. When he arrived at SPEC, his dosimeter showed that he had received a dose of 1.6 millisievert during the 10 minutes drive. This made him realise that his choice of efficiency over thoroughness had been wrong and that the abnormal indication on the meter was right: the needle was not stuck, but had actually showed a dangerously high level of radiation. It is perhaps a coincidence that this happened on Wednesday, 2 January, two days after the celebration of the New Year.

London Bombings

On 7 July 2005 three explosions occurred at around 08:50 on the London Underground system: the first on the Circle line between Aldgate and Liverpool Street, the next at Edgware Road station and the third on the Piccadilly line between Russell Square and King's Cross. At 09:47 a fourth explosion occurred on the upper deck of a London bus in Tavistock Place. Fifty-two people were killed and several hundred injured.

In the aftermath of a terrorist attack such as this, the immediate question is whether it could have been prevented. This quickly becomes a question of whether the authorities, in this case MI5, could have known or should have known that a real risk existed. When the identities of the four suicide bombers first became known, senior police and government officials maintained that the bombers were men who had never crossed the radar of the security service or Scotland Yard. The official view was that no intelligence had been missed and that the attacks 'came out of the blue.' Nobody could have foreseen the attacks, nobody slipped up and nobody could possibly bear any blame other than the bombers themselves. However, after a while it became publicly known that the official position was not actually correct. The Intelligence and Security Committee (ISC) was therefore asked to look into the case, and published a report with the result of their investigation in May 2006.

The intention here is not to review the case or to argue whether someone knew, did not know, or should have known. The press does a much better job of that as, e.g., shown by a headline in *The Guardian* (1 May 2007) which bluntly stated that 'MI5 decided to stop watching two suicide bombers.' The intention is rather to show that even here one can see the effects of an efficiency-thoroughness trade-off. Two comments from the ISC report are noteworthy in this respect:

> 48. As for the meetings in 2004, we found that they were covered by the Security Service as part of an important and substantial ongoing investigation. Siddeque Khan and Shazad Tanweer were among a number of unidentified men at the meetings. The Security Service did not seek to investigate or identify them at the time although we have been told that it would probably have been possible to do so had the decision been taken. The judgement was made (correctly with hindsight) that they were peripheral to the main investigation and there was no intelligence to suggest they were interested in planning an attack against the UK. Intelligence at the time suggested that their focus was training and insurgency operations in Pakistan and schemes to defraud financial institutions. As such, there was no reason to divert resources away from other higher priorities, which included investigations into attack planning against the UK.

> 55. It is also clear that, prior to the 7 July attacks, the Security Service had come across Siddeque Khan and Shazad Tanweer on the peripheries of other surveillance and investigative operations. At that time their identities were unknown to the Security Service and there was no appreciation of their subsequent significance. As there were more pressing priorities at the time, including the need to disrupt known plans to attack the UK, it was decided not to investigate them further or seek to identify them. When resources became available, attempts were made to find out more about these two and other peripheral contacts, but these resources were soon diverted back to what were considered to be higher investigative priorities.

The two excerpts show that investigators and administrators several times had to make a choice between efficiency and thoroughness. In this case efficiency meant focusing on the cases (or people) that seemed to represent the largest risk, given that limitations in manpower, time and resources made it impossible to focus on every single suspect. Thoroughness on the other hand meant just that, looking at every suspect and every possibility and following all leads until it was reasonably certain – again an ETTO-like choice! – that they did not lead anywhere. In every case where a choice was necessary, a choice was obviously made – and that choice itself may sometimes have been routine and sometimes more deliberate, hence also subject to the ETTO principle.

The resources available for counter-intelligence and surveillance are always limited (cf. below), and it is therefore necessary in practice to prioritise searches and investigations. An authority, such as the MI5, should not be blamed for doing so even in cases where it turns out that the wrong choice was made. It stands to reason that *if* the adverse outcome could have been foreseen, *then* the corresponding choice would have been different. But the very conditions that made a choice necessary also forced the choice to be a practical (and hence an approximate) rather than a 'rational' decision. Even on this level of what clearly must be called administrative decision-making, we recognise the ETTO fallacy, i.e., that people were required to be both efficient and thorough at the same time – or rather to be thorough when with hindsight it was wrong to be efficient!

Full Protection not Feasible

Resource limitations are often a problem when we try to prevent something or try to establish protection. One of the most conspicuous examples is the prevention against acts of terror, which most people in the world experience when they have to board an aircraft to travel somewhere. The use of metal detectors to screen passengers for weapons became common in the early 1970s after a series of high-jackings. Today more sophisticated detectors and methods are in use, often making the entry to the clean area of the airport a lengthy and slow procedure. Yet the use of such barriers to prevent certain classes of adverse events from taking place is itself subject to the ETTO principle, both on the level of the people operating the system (screeners) and on the level of managing the system. (As regards the former, it is a sobering thought that tests by the US Transport Security Administration in 2007 showed that roughly 75 per cent of fake bombs, or component parts to bombs, were missed by screeners. In this case the number of people waiting to pass the control point clearly limits the time that can be spent on screening each person. So much for thoroughness!)

In the so-called 'war against terror,' Homeland Security Secretary Michael Chertoff in 2006 told a Senate committee that full protection was not feasible. To protect against every possible threat would require billions of dollars in spending in order to finish installation of radiation detection equipment at ports; to build fences or high-technology barriers at borders to control illegal immigration; to enhance railroad safety programmes; to install new explosives detection equipment at airports; and to inspect every container shipped into the US. It would, as Chertoff said, 'bleed America to the point of bankruptcy.'

To take a simple example, it is impossible to screen every passenger boarding a train, a bus, or an underground train system. The risk of such a person carrying explosives is real, as too many unfortunate events have shown. And it is probably higher than the risk of an airline passenger carrying a bomb – not least because the latter know that they will be searched. But quite apart from the cost of screening everyone boarding a means of public transportation, it would grind the whole system to a halt. Full protection is not feasible both because it would cost too much and because the efficiency of the function(s) being protected would be driven towards zero.

In another domain, full protection against traffic accidents would, for instance, mean that the maximum speed should be 20 kilometres per hour, or even lower. This would not eliminate collisions, since only a complete ban on driving could do that. But it would reduce the speed to a level where the commonplace safety devices (at least in newer cars), such as safety belts, airbags, and active braking systems, would ensure that people in the car would be safe. (For people outside the car, such as pedestrians and cyclists, it is a quite different story.) Yet it does not take much imagination to realise what this would mean for efficiency, for instance such a simple thing as getting to work – and getting back again. Add to that the distribution of goods, services, etc., and the picture is clear. Full protection is not feasible because it would be counter to the aims of an effective society. It is also impossible because it is too expensive, because it is impossible to think of everything that could go wrong or constitute a risk, etc.

The Challenges of Publishing Aeronautical Data

One prerequisite for safety is knowledge of what has happened in the past. In many fields of activity this has become a *sine qua non*, a requirement from authorities and the public, and consequently a service or function in its own right. Providing an analysis and a summary of events that have taken place must be done accurately, frequently, and quickly. It must be accurate because the information will be used to make decisions about safety, resource allocations, operational conditions, etc. It must be frequent so that it reflects the relevant past and is up to date. And it must be quick so that the feedback is not delayed or comes too late to be of any use. In many organisations, providing summaries of past events is institutionalised, with regular reporting periods and with requirements to publish results with little or no delay. There is, in other words, a demand for thoroughness as well as a demand for efficiency. It is therefore no wonder that many examples of the ETTO principle can be found in this field of activity.

In aviation, the International Civil Aviation Organization (ICAO) requires member organisations to collect and distribute areonautical performance data to domestic and international users on a weekly basis. For any given country that consists in collecting the data, performing content and format editing and publishing the material in a weekly document.

In Canada, the current publication process has been in place for more than a decade but has problems in meeting today's increased demands and technical requirements, due to the sheer volume of work, the demands for precision and the limited time frame. Demands and constraints arise from legislated periodic reporting deadlines, relationships between system partners, and the highly regulated nature of the aviation business. The problems are accentuated by an obsolete data network with insufficient capacity and dysfunctional procedures that create information bottlenecks, where higher pressures and volumes of information lead to increased error rates.

The overall process is shown in Figure 4.2 below. The Regional Offices verify aeronautical data for correctness based on publication requirements. After checks have been performed, the Proposed

Aeronautical Changes (PAC) are forwarded to the Head Office where specialists re-evaluate the data to ensure it meets procedural rules and checks the PAC against a plethora of data including frequencies, aerodromes, obstacles, navigational aids, and instrument procedures. At this stage, the PAC is converted into a Subject Data Form (SDF) and added to a database used by three different groups to create their unique products: Natural Resources Canada (NRCan), Nav Canada Aeronautical Information Database (NAIDS), and the Aeronautical Information, Regulation and Control (AIRAC).

Work in the AIRAC office is driven by the need to produce a weekly report that is as accurate as possible. Despite resident quality assurance programmes and an ISO certified in-house reporting system, the demands to turn raw aeronautical data into manageable, legible bits for consumption both at home and abroad means that operators and managers must make many efficiency-thoroughness trade-offs. A few examples will suffice.

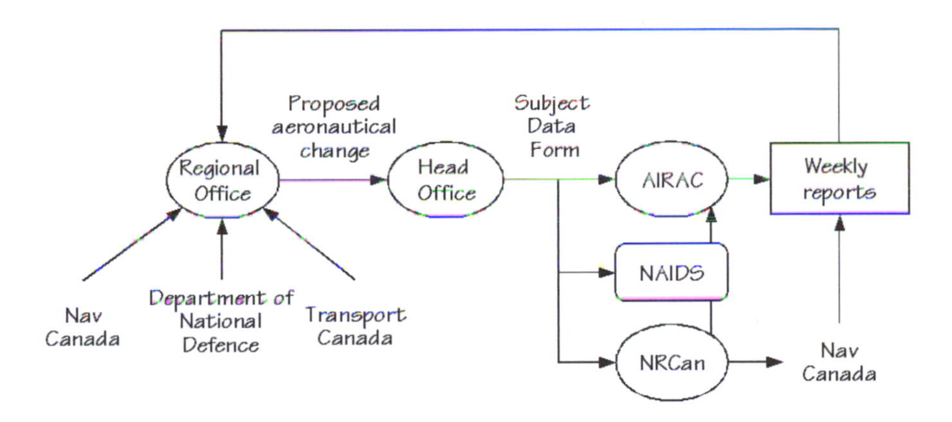

Figure 4.2: The AIRAC production flow (simplified)

- *Looks OK, it has been checked earlier by someone else.* Because of time constraints and production pressures, AIRAC editors are never able to verify every SDF to the full extent required. Should some piece of background information be missing, it is impossible to track each historical element linked to a particular piece of data. While the AIRAC is responsible for the work that is published, it is still necessary to assume that the operator who processed the item beforehand made double and triple checks for correctness. Instead of AIRAC editors conducting a full verification, the onus is on the contributor to get the facts right.

- *Normally OK, no need to check it now.* A final check of an AIRAC document is rarely deferred, although the need may sometimes arise on the last day of a cycle where the issue is very complex, and due to content, awkward in size. As an example, 'significant obstacle tables' are normally straightforward with little chance for anomaly. The editors may therefore ignore their presence in the AIRAC document and feel confident of the tables' integrity level. If for some reason there was an error, the editors would conduct a review and reverse the statement in the next edition, effective on the next cycle date of information.
- *Insufficient time and resources, will do it later.* While it is difficult if not impossible to defer a task within the primary publishing duties, non-AIRAC tasks are often pushed aside. Examples are creating designators for new navigational aids and the ever-abundant customer service work such as processing ICAO call-signs and designator applications. Yet postponing the non-AIRAC related work in a timely fashion is only a temporary solution. There comes a time in every cycle where the editors must attend to every aspect of their job, despite the overriding pressure in creating a weekly AIRAC issue.

Several other ETTO rules can also be recognised in this work, such as the unfortunate combination of 'it will be checked later by someone else' and 'it has been checked earlier by someone else.' In a concerted effort, AIRAC editors have successfully developed a set of *ad hoc* in-house procedures that do buy them more time. Despite the considerable production pressures, they are able to maintain the quality of their work. This case therefore illustrates how the ETTO rules can be used to develop an efficient way of working without creating unacceptable risks or hazards.

Incorrect Chemotherapy Treatment

This case is about a 67-year-old semi-retired cattle farmer who had been experiencing fatigue and shortness of breath for a few months, but who had delayed visiting his family physician until the calving season was completed. Following an initial examination, the patient was referred to the nearest community hospital, where he was diagnosed as suffering from a primary carcinoma of the stomach (gastric carcinoma) which had spread to the lungs (metastases). The appropriate chemotherapy was ordered and treatment started soon after. Parallel to that, a biopsy was analysed by two pathologists who after some time concluded that the underlying pathology was a lymphoma, which is quite different from a gastric carcinoma. Unfortunately the report was missed and the incorrect chemotherapy was started. Almost five months later the patient died.

An analysis of this tragic event reveals ETTOing at all levels. The patient was for instance assessed in the oncology clinic on a day when 35 other patients were being seen by the same physician. The pressure from the hospital administration to maintain a high volume of patients (euphemistically called 'service to the community') had become normal but conflicted with the ideal to provide safe service, thus creating a double-bind situation for the professional staff. The large number of patients, which by the way was typical for this clinic, also created a bias towards efficiency in the examination of patients.

After the first diagnosis had been made, steps were taken to start the chemotherapy without waiting for the final pathology report. In this case both doctor and patient were under time pressure. The patient wanted to start treatment as soon as possible because of an impending family wedding and because the haying season would begin within a month. The doctor wanted to get the chemotherapy prepared because of going to an international conference that would last for two weeks. If the doctor waited until after returning, the treatment might not get underway before the start of the haying season. The insufficient time that both patient and doctor experienced interacted in an unusual and unfortunate way, and made both prioritise efficiency and speed over thoroughness.

A third ETTO case was the lack of verification before beginning the chemotherapy. This should have been done, even when the chemotherapy had already been ordered (cf. above). The final pathology report did in fact arrive in the oncology clinic before the chemotherapy was started, but nobody noticed it. One reason may have been that the report was types in a small font on a form in which the surgeon's pre-procedure diagnosis and post-procedure diagnosis were clearly indicated in large bold font lettering. The final report was also stapled behind the initial report. Even if someone had tried to verify the diagnosis of the first report (gastric carcinoma), they would probably not have read all the material anyway (justified by 'no time,' 'it looks fine,' or 'it has been checked earlier by someone else'). The pathologist also sacrificed thoroughness for efficiency by only sending the biopsy report to the person who did the procedure ('do not use too many resources').

In addition to these, and other examples, of ETTOing, there also seemed to be a lack of proper procedures for verification of the pathology report as part of the several cycles of chemotherapy – even after the patient became profoundly anaemic. Procedures are, of course, no guarantee that people will be thorough rather than efficient, since procedure-following itself can be sacrificed in the name of efficiency and productivity. But thoroughness is even less likely in the absence of procedures, since there is nothing to remind people what they ought to do. The overall picture is that of a complex socio-technical system full of multiple, simultaneous goal conflicts – which by the way is not an unusual condition. Attempts to resolve these conflicts by trading off thoroughness against efficiency generate changes or variations in the conditions of normal work, which in turn affect the likelihood of an accident or a serious mistake occurring within that system.

A Reverse Fuel Situation

This case study concerns a company that operated a Douglas DC-3 for vintage charters and scenic flights. The DC-3 is a fixed-wing propeller driven aircraft, which first flew on 17 December 1935. Douglas built more than 10,000 planes of this type, which remained in use well into the 1970s, and is still flown for special purposes. The company discussed here had been doing scenic flights for a number of years and was licensed and approved by the Civil Aviation Administration. The technicians as well as the aircrew were also licensed and experienced.

The standard operation procedures state that the main fuel tanks always must be filled with the required fuel (reserve fuel included) for the flight and the auxiliary fuel tanks must have 45-minute reserve fuel. Due to the inaccuracy of the fuel gauges, the fuel tanks must be dipped before the flight to determine the correct fuel levels.

The flight for the day was a ferry flight from base to a nearby general aviation airport followed by a 1-hour scenic flight with 15 passengers, returning to the general aviation airport, and then a ferry flight back to base. On the previous day, the aircraft had returned from another flight where it had been refuelled totally and then used both the main and auxiliary fuel tanks. On shut-down the auxiliary tanks held the required fuel for the next day and the main tanks held approximately 1 hour's reserve. On the day of the flight the aircraft was prepared and the fuel tanks dipped by the senior technician, but the information about the reverse fuel situation was not passed on.

During the briefing of the aircrew, the fuel requirement was confirmed by flight operations who knew about the reversed fuel situation but who failed to highlight the potential problem. During the pre-flight checks the captain directly asked a technician if the main fuel tanks had been dipped for the correct fuel requirement, and received an affirmative answer. On start up the main fuel tanks were selected since nobody at this stage had mentioned the reverse fuel situation.

Both the ferry flight and the scenic flight went off without any problems. Shortly after take-off for the return flight to base the left-hand engine began to run rough. The first officer was pilot flying and continued to do so as the captain took charge of the emergency. The left-hand auxiliary tank was selected as the right-hand engine began to run rough. While continuously selecting between auxiliary and main tanks left and right the captain could not restart either engine. The first officer carried out a successful forced landing and there were no injuries.

The ETTO principle can be used to understand how this situation came about. Even a quick look will show examples of ETTOing on all levels of the organisation.

- *Negative reporting.* The senior technician did not pass on the information of the reversed fuel situation as there was more than sufficient fuel on board for a safe flight. The pilots therefore worked on the assumption that the conditions were as normal.
- *Management double standard.* The aircraft was not grounded due to inaccurate fuel gauges but a safety policy was introduced whereby the fuel tanks would be dipped before a flight. While the official policy was safety first, in practice efficiency came first.
- *It has been checked earlier by someone else.* The aircrew knew that the fuel had been checked earlier so they did not dip the fuel every time they landed that day.

- *Normally OK, no need to check.* The aircrew did not notice any discrepancies between the fuel tank indications due to the accepted practice of flying with inaccurate fuel gauges.
- *We always do it this way here.* The aircraft had been operating for some time with the inaccurate fuel gauges and the fuel dipping safety policy. The aircrew knew this was now standard practice for such operations.
- *Not really important.* 'It looks fishy, but I don't think the consequences are really serious.' The aircrew knew about the inaccurate fuel gauges but also knew about the fuel dipping safety policy.

The reverse fuel situation created a problem for the pilots because they did not know about it. Fortunately, the outcome was not fatal. This case is yet another illustration of the fact that local adjustments – through shortcuts, heuristics and expectation-driven actions – are the norm rather than the exception on both the individual and organisational levels. Normal actions are successful because people learn how to adjust to the local conditions, including the adjustments made by others. The ETTO rules can be seen as the – reasonable – assumptions that people make about the situation at work. Because these assumptions are normally fulfilled, they also learn that it is unnecessary to check them. This may every now and then cause surprises, sometimes benign, as in this case, but sometimes unfortunately also very serious.

The Loss of M/V *Estonia*

The M/V *Estonia* was a car and passenger ferry built in 1979. It was originally named *Viking Sally* and was put into service on the route between Turku, Mariehamn and Stockholm. In 1993 she started to sail on the Tallinn–Stockholm route under the name *Estonia*. She sank about 01:50 on 28 September 1994 in the Baltic Sea, taking 852 lives. Remote videotapes of the wreck showed that the locks on the bow visor had failed and that the door had separated from the rest of the vessel. The accident was eerily reminiscent of the disaster that happened on 6 March 1987 when the *Herald of Free Enterprise* capsized in the English Channel just outside the Belgian port of Bruges-Zeebrugge with the loss of 193 lives. Both vessels were roll-on roll-off car and passenger ferries, and in both cases a large quantity of water entered the flat car deck, leading to a loss of stability.

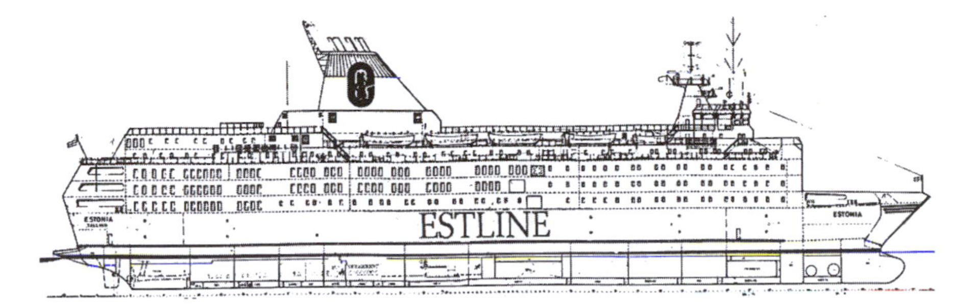

Figure 4.3: Estonia, *showing the position of the car deck*

Both disasters have been analysed at length, and both show clear examples of ETTOing. In the case of the M/V *Estonia*, the effects of the ETTO principle can be seen in the very construction of the vessel (Figure 4.3). The design of a roll-on roll-off ferry represents a fundamental trade-off, where the ability quickly to get cars onto and from the car deck is more important than protecting against known risks, for instance by dividing the car deck into smaller parts protected by bulkheads. During the building of the vessel details of the forward construction, consisting of the hinges and lockings to the bow visor and the ramp, were never controlled specifically. The classification society had no specific demands on these details, hence did not consider it as part of their responsibility, while the Finnish authority regarded it as something to be taken care of by the classification society. Both organisations thus thought that the arrangement would be checked by somebody else, a clear ETTO rule.

The weather on the night of the accident was quite bad with a wind of 15 to 20 m/s and a significant wave height of 3 to 4 meters. M/V *Estonia* apparently did not slow down but tried to sail at normal speed; when the captain was first notified of the heavy blows by the waves on the bow (around 00:58), he only asked how many engines were running and commented that they were already one hour late. This strongly indicates an efficiency-thoroughness trade-off, where maintaining a regular voyage schedule was the overriding concern. Around 01:00 some slamming was heard and the officer of the watch sent the seaman of the watch to check what it was all about. The seaman stopped on the way to chat with some crew members at the reception desk and failed to reach the vehicle deck in time. Meanwhile, at about 01:15 the bow visor became loose. As it was hung on the upper part of the ramp, connected via the recess construction, the visor pulled the ramp causing it to loosen and consequently it opened. As the ramp opened water started to flood in and soon the ship listed. Shortly after, it sank.

This accident illustrates how a number of latent conditions (car deck design and construction of bow visor and ramp) interacted with the actual conditions and events (the strong wind and the failure to reduce speed) in a very unfortunate manner. The ETTO principle is a help to understand how the latent conditions arose and how the hazardous situation came about, and furthermore shows how failures at the sharp and blunt ends have the same aetiology. This should really not be surprising, since people's performance must be governed by the same principles no matter where it takes place.

Avoiding Collision at Sea

Whenever two or more vehicles move in the vicinity of each other, on land, at sea, or in the air, there is a possibility of a collision. (Even pedestrians can, of course, bump into each other under the right – or rather, wrong – conditions.) In the case of maritime traffic, a set of rules called *The International Regulations for Preventing Collisions at Sea 1972* (*COLREGS*) has been devised to make sure that collisions do not take place under normal conditions. In principle, if all vessels follow the rules, collisions should not happen. The fact that collisions do happen therefore strongly suggest that the rules sometimes may not be followed. One reason for that is this navigators, ships' masters, first and second officers, like everyone else are prone to make efficiency-thoroughness trade-offs. A closer look at some of the rules provides an idea about how this may happen.

Each *COLREGS* rule by itself is brief and straightforward to understand. For crossing vessels in sight of one another, Figure 4.4 illustrates the three main rules:

- *Rule 15. Crossing situations.* When two power-driven vessels are crossing so as to involve risk of collision, the vessel which has the other on her own starboard side shall keep out of the way and shall, if the circumstances of the case admit, avoid crossing ahead of the other vessel.

- *Rule 16. The give-way vessel.* Every vessel which is directed to keep out of the way of another vessel shall, so far as possible, take early and substantial action to keep well clear.
- *Rule 17. The stand-on vessel.* The stand-on vessel may take action to avoid collision if it becomes clear that the give-way vessel is not taking appropriate action.

In Figure 4.4, vessel A is the non-privileged or 'give-way' vessel, because it has vessel B, the privileged or 'stand-on' vessel to starboard. Depending on the relative speed and course of the two vessels, vessel A can decide either to maintain the course but slow down so that it arrives at point X after vessel B, or to change course to starboard in order to sail behind vessel B. According to Rule 15, vessel A should avoid crossing ahead of vessel B, since that clearly can create a risk. (In practice, the master of a modern vessel can make use of navigational aids to determine both the Distance at Closest Point of Approach (CPA) and the Time to Closest Point of Approach (TCPA), and use this information to decide which action to take.)

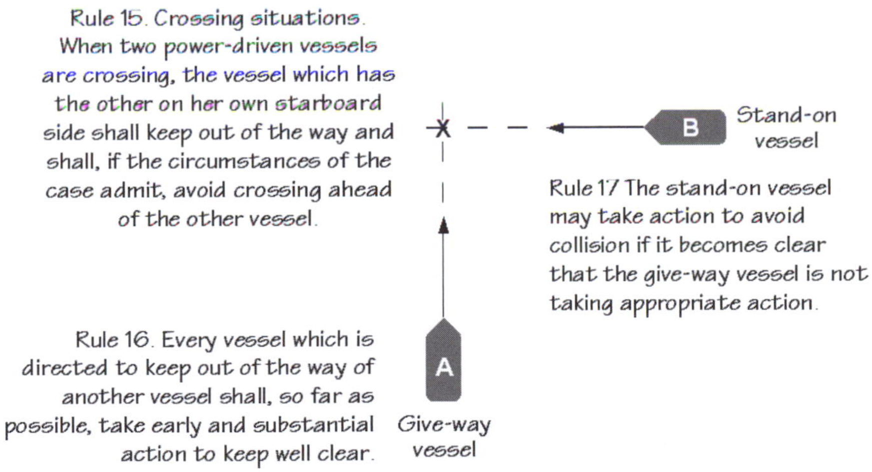

Figure 4.4: Three COLREGS rules

In narrow waters a set of rules for avoiding collisions are necessary because the intensity of traffic can be very high. Consider, for instance, Dover Strait, the waterway that separates England and the European Continent. Four hundred vessels pass through Dover Strait each day, with an additional 70 ferries and 240 other vessels crossing from side to side. (There is so much traffic in the Channel that in 1967 a separation scheme was introduced so that vessels travelling south-west from the North Sea sail on the British side of the Channel and vessels travelling in the opposite direction steer closer to the French side.) A ferry on its way from Calais to Dover, for instance, will regularly have to resort to the collision rules, and a practice will therefore soon be established. Thoroughness, of course, means following the *COLREGS*. As the example above has shown, this either means that the vessel must slow down or change course. Both options are likely to increase the time it takes to cross from one side of the Channel to the other, which for a ferry is something to be avoided. From an efficiency point of view it may therefore be better for vessel A to cross ahead of vessel B, even though it breaks Rule 15. Studies of ferries crossing Dover Strait indicate that in situations where a ferry is the give-way vessel, it will in about 20 per cent of cases change course to port rather than to starboard, and thereby break Rule 15. Even if the situation in practice is more complex than the example shown in Figure 4.4, due to the presence of other vessels, differences in speed, late or ambiguous movements of the other vessel(s), etc., the benefit of making these trade-offs clearly outweighs the costs, which is why they become part of established practice. 'Sailing to rule' might arguably be safer, but it will on the whole also be less efficient. (For those of us who are not ships' masters, analogous situations often occur while driving in a city, although the rules are less strictly formulated in this case.)

Mission to Mars

The ETTO principle describes a seemingly universal trait of human performance, regardless of whether the scope is managing the situation at hand or making a strategic decision. It is therefore not surprising that we can find the ETTO principle even in matters concerning the exploration of space.

The dream of travelling to the moon, the planets, and the stars is as old as mankind itself. In modern times, the dream has become a reality thanks to a number of technological developments – in materials, in propulsion techniques, and in computing. One of the dreams was to send a spacecraft to Mars, to find out more about the mysterious red planet. This dream came true on 20 July 1976 when the *Viking 1* lander touched down on the surface of the planet, followed shortly after, on 3 September, by the *Viking 2* lander. Both landers had been launched from Earth almost a year before and had taken 11 and 12 months to reach their destination, respectively.

One part of the preparations for this mission to Mars was to consider the possibility of contaminating Mars with organisms from Earth. This possibility had been foreseen in the 1967 international 'Treaty on Principles Governing the Activities of States in the Exploration and Use of Outer Space, Including the Moon and Other Celestial Bodies,' in which Article IX asserts that:

> 'States Parties to the Treaty shall pursue studies of outer space, including the moon and other celestial bodies, and conduct exploration of them so as to avoid their harmful contamination and also adverse changes in the environment of the Earth resulting from the introduction of extraterrestrial matter ...'

The concern was raised by the US Congress and the risk was taken very seriously. Even though there was no empirical basis for determining what the probability of growth of imported terrestrial microbes on Mars might be, it was in the end agreed that the acceptable risk should be less than one billionth ($p < 10^{-9}$). (The reader is invited to consider the reasonableness of this value.)

In order to meet this criterion, each lander was covered with an aeroshell heatshield from launch until Martian atmospheric entry, to prevent contamination of the Martian surface with Earthly microbial life. As a further precaution, each lander, upon assembly and enclosure within the aeroshell, was 'baked' at a temperature of +250 °F (120 °C) for a total of seven days, after which a 'bioshield' was placed over the aeroshell and only jettisoned after the upper stage of the launcher fired the *Viking* orbiter/lander combination out of Earth orbit.

While this method was thorough, and probably as thorough as the knowledge and technology of the times could support, it was also very complicated, time consuming, and costly. The method furthermore introduced the risk that the 'baking' could damage the sensitive instruments on board. So when data from the Viking landers showed that Mars was too dry, too cold, had too much UV radiation and too much CO_2 for Earth organisms to survive, there was less necessity to use the same procedure for future missions. As a result, the rules were relaxed so that only those parts of a rover that would dig into the soil or drill into rock had to be sterilised. The rest of the craft was allowed to carry up to 300 spores per square metre. This was clearly a trade-off between efficiency and thoroughness, but one that was justified according to the interpretation of the data at the time.

Later expeditions to Mars, most recently the *Phoenix* mission that landed on Mars on 25 May 2008, have unfortunately found that water does exist, and that the former strict precautions therefore were necessary after all. Water could rejuvenate any terrestrial spores on the rovers, particularly if the spores are ultra-hardy extremophiles, leading to contamination of the planet. NASA is therefore looking for new ways of sterilising the rovers, for instance by using an electron beam to kill any bugs. This renewed emphasis on thoroughness may, however, come too late. Mars may already have been contaminated by either Russian or US spacecraft that crash-landed, or the 1997 *Pathfinder* and the 2004 rovers that were minimally sterilised. The lesson to be learned is that for such high consequence endeavours, thoroughness should not be traded off too easily or too early.

Concluding Remarks

The examples described in this chapter have shown how the ETTO principle can be used to describe what people normally do and thereby also understand how adverse outcomes can be a result of normal performance. Many other examples could have been mentioned, both spectacular and mundane.

Most of the examples refer to situations where the outcome was unwanted, in the worst cases involving the loss of human life. The reason for this is that studies of failures are far more frequent than studies of successes. It does not mean that ETTOing only happens when things go wrong. In the case of the publication of aeronautical data, for example, the ETTO principle makes it clear how the system works and why it is able to succeed even though conditions are far from ideal. Even in the cases where something went wrong – such as the problems with aircraft engines in the winter, the incorrect chemotherapy treatment, or the loss of M/V *Estonia* – it is important to understand that the normal functioning of these systems – flights leaving and arriving in time, patients being diagnosed and treated, and ferries keeping their schedules – depends on regularly making the same performance adjustments in the same way. Something can therefore go wrong even if nothing fails!

It is also worth pointing out that the bias towards investigating failures rather than success itself represents a trade-off. The fact that time and resources are limited certainly means that it is impossible to study more than a subset of all possible cases. But this limitation does not mean that the subset should only contain failures. From a thoroughness perspective it would make a lot of sense to study successes in order to understand how these happen, even if the concern is with safety rather than business. Even if the probability of failure is as high as 10^{-4}, there are still 9,999 successes for every failure, hence a much better basis for learning. The ETTO thinking, however, means that successes are looked upon as '*not really important*' and '*it is normally OK*,' and also fall prey to the rule of '*negative reporting*.' Efficiency therefore seems to justify that only events with adverse outcomes – in many cases reduced further to be only events with serious adverse outcomes – are put under scrutiny. While this may pay-off in the short term, the long term value of this attitude is in serious doubt.

Sources for Chapter 4

Of the nine examples described in this chapter, five make use of data and information that is freely available in the public domain. For some of them, such as the Studsvik transportation case or the London bombings, official investigation reports can also be found on the internet. A source for the Channel ferry example is C. Chauvin and S. Lardjane, (2008), 'Decision making and strategies in an interaction situation: Collision avoidance at sea.' *Transportation Research Part F*, 11, 259–269.

The remaining four examples are taken from an exercise in ETTO analysis that was part of a master's course in October 2006 at the University of Lund, Sweden. The analyses were in each case done by subject matter specialists and the abbreviated versions used in this chapter fail to do justice to the richness of the original analyses. I am nevertheless grateful for having been permitted to use excerpts from the following reports in this chapter:

- 'The challenges linked to publishing Canadian aeronautical data,' Dave Mohan, AIRAC Editor, Aeronautical Information Services and Flight Operations, NAV CANADA.
- 'Incorrect chemotherapy treatment resulting in patient death,' Rob Robson, Chief Patient Safety Officer, Winnipeg Regional Health Authority, Canada.
- 'A scenic flight arrives at the scene,' Mike Weingartz, Senior Maintenance Test Pilot, Denel Aviation, South Africa.
- 'The loss of M/V *Estonia*,' Jörgen Zachau, Captain, Swedish Maritime Administration.

Chapter 5: The Usefulness of Performance Variability

'Nearly all hazardous operations involve making actions that lie outside the prescribed boundaries, yet remain within the limits of what would be judged as acceptable practice by people sharing comparable skills' (Reason, 1997, p. 51)

When a hazardous operation has resulted in an adverse outcome and therefore is investigated, it is practically always found that people have acted differently from what they were supposed to or expected to. The essence of the above quote is nevertheless that it is wrong universally to label such performance variability as being incorrect, e.g., as a 'human error' or a violation. The reason is that performance variability or 'actions outside the prescribed boundaries' happens all the time. In hazardous operations the performance variability may sometimes lead to unwanted outcomes and it is therefore noticed. But the very reason for noticing it biases both *what* is noticed and *how* it is interpreted. Performance variability is clearly not restricted to hazardous operations, but is found in all kinds of human activity and at all levels of work.

In accident investigation, as in most other human endeavours, we fall prey to the *What-You-Look-For-Is-What-You-Find* or *WYLFIWYF* principle. This is a simple recognition of the fact that assumptions about what we are going to see (*What-You-Look-For*), to a large extent will determine what we actually find (*What-You-Find*). (The principle is furthermore not limited to accident investigation, but applies to human perception and cognition in general.) In accident investigations, the guiding assumptions are sometimes explicit, for instance when they are given as a directive or objective. (A recent example of that is the explosion at BP's Texas City refinery on 23 March 23 2005, in which 15 people were killed and more than 180 injured. In this case a number of investigations were carried out, each having different premises or instructions, and each therefore coming to different conclusions. In one investigation, the team was instructed generally 'to limit their efforts to the conditions and circumstances leading to the incident, and avoid doing a general safety audit'!) But in most cases the assumptions are implied by the methods that are used. As an example, a Root Cause Analysis takes for granted that accidents can be explained by finding the root – or real – causes, and that the accident can be described as a sequence, or tree, of causes and effects. Other methods may have names that are less evocative, but all have implied assumptions nevertheless. Or as Thomas Hobbes wrote, in 1651, in *Leviathan*: 'Ignorance of remote causes, disposeth men to attribute all events, to the causes immediate, and Instrumentall: For these are all the causes they perceive.' Since we can never find what we do *not* look for, serendipity excepted, our initial assumptions about possible causes will invariably constrain the investigation.

The investigation bias obscures the fact that there are very few situations where things go wrong and many, many more where things work out fine and where the outcomes are as intended, as expected, or at least acceptable under the circumstances. Yet when outcomes are acceptable there is little motivation to look into the reason for that; it is simply taken for granted – and even considered normal – that things go right. Performance variability is consequently not noticed, even though it can be found under these conditions as well. One argument for why this must be so is simply that it would be unreasonable to assume that people, or organisations, behaved in one way when things went wrong and in another when they went right. 'Right' and 'wrong' are judgements made after the fact, and it would be miraculous, to say the least, if the 'machinery of the mind' could know the actual outcome of an action ahead of time and change its mode of functioning accordingly.

Underspecified Systems

The view that performance variability is useful is consistent with the principles behind Charles Perrow's *Normal Accident Theory* that were mentioned in Chapter 1. Perrow argued that by the beginning of the 1980s many socio-technical systems had become so complex that they could no longer be fully controlled, and that accidents therefore should be treated as normal rather than as exceptional occurrences.

Perrow proposed that systems in general could be described by two dimensions called *coupling* and *interactiveness*, respectively (see Figure 5.1). Coupling describes the degree to which subsystems, functions, and components are connected or depend upon each other; the degree of coupling can range from *loose* to *tight*. Interactiveness describes the degree to which events in the system develop in ways that are expected, familiar, and visible, or whether they develop in ways that are unexpected, unfamiliar, and invisible. The degree of interactiveness can range from *linear* to *complex* interactions, respectively.

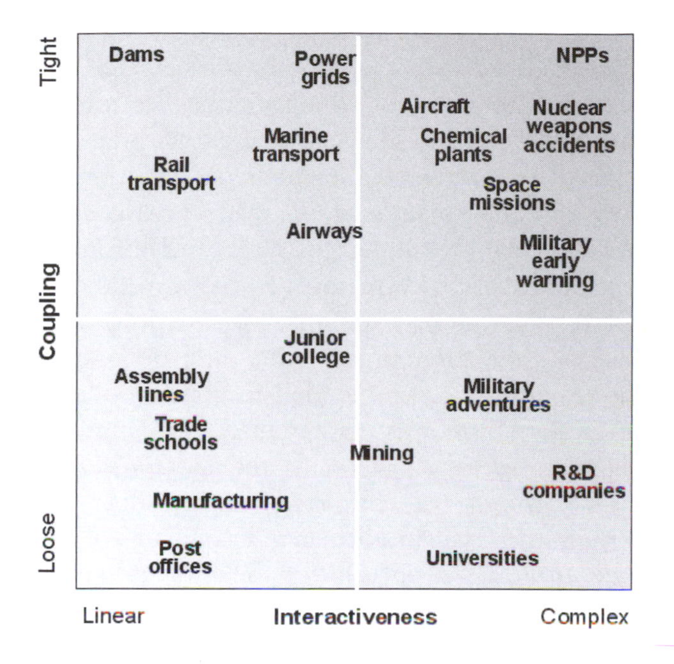

Figure 5.1: Coupling–interactiveness relations (after Perrow, 1984)

Figure 5.1 shows how these dimensions were used to characterise different types of socio-technical systems. According to the reasoning behind the dimensions, accidents were 'normal' in systems characterised by complex interactiveness and tight couplings, i.e., systems in the upper right quadrant. Because the interactiveness was complex, it would be difficult for the people in charge of the system to understand what was going on; and because the couplings were tight, the interventions required to direct or control the process would have to be correct and precise, with little or no room for variability.

For the present discussion, the possible relation between the controllability of a system and the coupling–interactiveness dimensions is more important than how systems are rated on the dimensions as such. In the case of coupling, systems that are tightly coupled require more precise control actions or interventions than systems that are loosely coupled. In tightly coupled systems, such as power grids and nuclear power plants, delays in processing are not possible, sequences of actions cannot be changed, and there is usually only one way to reach the goal. In tightly coupled systems, a change in one part of the system will quickly spread to other parts. In loosely coupled systems, such as post offices and universities, the opposite is the case. While this does not necessarily make them any easier to manage, their performance depends less on whether activities or functions are precisely controlled or aligned at every moment in time.

In the case of interactiveness, systems on the complex end of the scale are more difficult to understand than systems that have linear interactions. There may be indirect or inferential information sources, limited understanding of processes or the ways in which changes occur, many control parameters with potential interaction, many common-mode connections of components that are not in a production sequence, and unfamiliar or unintended feedback loops. It is thus the consequence of the complexity, rather than the complexity itself that matters. This may be easier to see if the *interactiveness* dimension is replaced by a *manageability* (or *controllability*) dimension, which also more directly describes the ease or difficulty by which something can be managed or controlled. On one end of the scale are systems with low manageability, and on the other systems with high manageability, cf. Figure 5.2. (The difference between Figure 5.1 and Figure 5.2 is not just that *interactiveness* has been replaced by *manageability*, but also that the systems that are used to illustrate the dimensions have been readjusted to reflect the situation in 2009.)

On a more aggregate level, perhaps the most important feature of a system is how tractable it is. A tractable system is easy to control, while an intractable system is difficult or even impossible to control. Many present-day systems, not least those that are of major interest for industrial safety (such as power generation, aviation, chemical and petrochemical production, healthcare, transportation, etc.) are intractable rather than tractable, cf. Chapter 3. This also goes for large-scale IT systems, such as the internet or radio networks, and for practically all socio-technical systems whether they are complex, as the ones mentioned above, or simple.

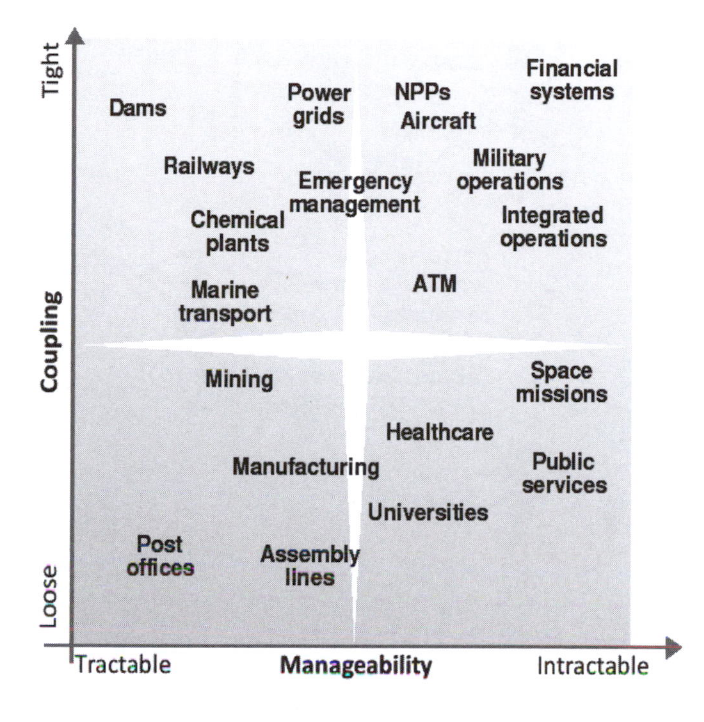

Figure 5.2: Coupling–manageability relations

The Need of Performance Variability

If a system is underspecified there will be situations for which it is impossible to provide a detailed description of tasks and activities. For such situations, adjustments or compromises must perforce be made when the actions are carried out; in other words, performance must be variable. In consequence of that, outcomes cannot be deterministic but must be probabilistic – although usually with a high degree of certainty. The underspecification will likely apply to most tasks and activities in the system. This means that performance of an activity *X* will be variable both because the system is underspecified, and because other activities (inputs, resources, instructions, etc.) are variable, hence the result may be unpredictable inputs, cf. Figure 5.3. A few examples will illustrate the universality of performance variability.

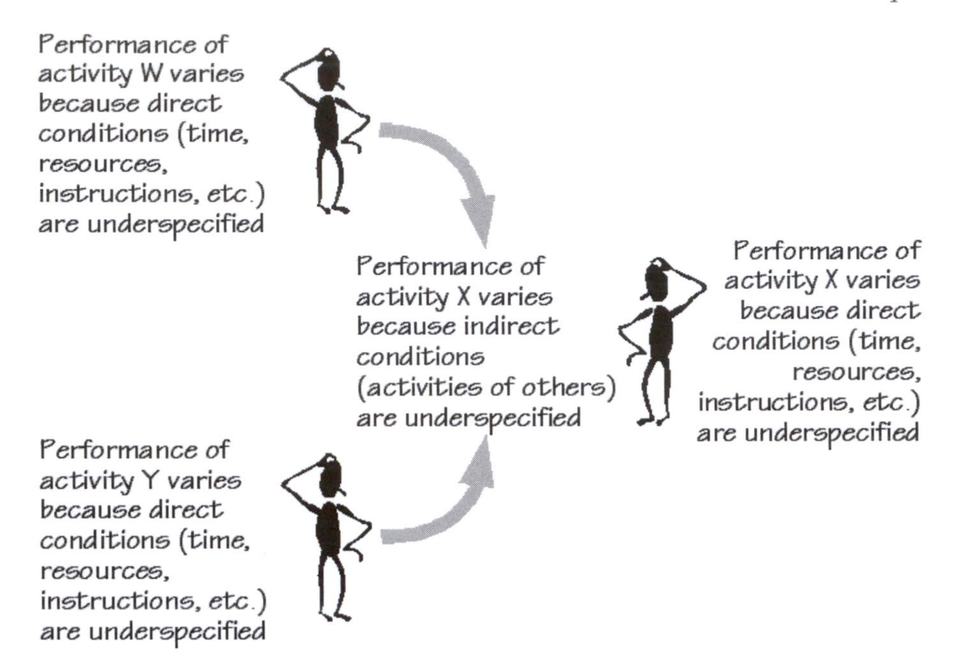

Figure 5.3: Direct and indirect sources of underspecification

- Consider a simple example such as taking a photocopy of a document. Here the environment is stable and predictable (with the exception of occasional paper jams), in the sense that the tasks are simple and well-defined and that the user therefore can focus on controlling the photocopier and getting the work done. It is also possible to prescribe precisely what to do, i.e., to give a step-by-step instruction. In this case there is little or no need for the user to adjust their performance, and doing so anyway will probably lead to a loss of efficiency.
- Consider another simple example, such as driving to work, which millions of people do every day. While it may be possible to provide an instruction about how to control the vehicle (accelerate, brake, steer), it is impossible to describe how the vehicle should be controlled as it moves through the traffic (e.g., when to stop and start, when to turn, etc.). The reason is that the environment is dynamic, hence intractable and unpredictable. So while a navigation system can tell you which road or route to take, a driver would be ill-advised blindly to follow its instructions and ignore the actual

traffic. The roads themselves do not change, but the traffic does. Driving requires a continuous adjustment of speed and direction to avoid collisions with other vehicles (or bicycles or pedestrians) and at the same time to ensure sufficient progress. Most drivers can do that so effectively that accidents are amazingly rare, considering the density (and speed) of traffic. In driving, performance variability is necessary and helps to maintain effective performance of the system (all the drivers, all the traffic).

- An industrial case would be something like work in a process plant, e.g., an isomerisation unit at a refinery. For such tasks a normal operating procedure is usually provided, and operators are on the whole expected to follow that, in spirit if not to the letter. There will, however, always be some variability because of the environment (temperature, humidity), state of equipment, time of day, etc. In cases where, e.g., a piece of equipment (a sensor, an actuator, a machine) malfunctions there are normally few procedures, which means that the situation is underspecified. (For nuclear power plants and other safety critical installations there may be so-called Emergency Operation Procedures. But these are intended to be used as guidelines rather than instructions.) The reason is simply that there are so many possible variations of what could happen, and so many unknown factors, that it is impossible to predict and describe exactly what the situation will be. It is therefore also impossible to provide procedures for it. If the situation is to be managed, it is necessary to adjust performance to meet the conditions, i.e., to have performance variability. The requirement strictly to follow the procedure is usually made *post hoc*, especially if something has gone wrong. In this case it serves mainly as a justification for apportioning blame.

Most systems normally have sufficient damping to ensure that the performance variabilities do not combine in a manner that may destabilise the situation. But every now and then the situation and the overall performance variability may escalate and eventually destabilise the system.

Sources and Types of Performance Variability

Performance variability is a unique and necessary feature of socio-technical systems, and more specifically of humans and of social entities (organisations), but is not found in machines and technology. Machines are designed to carry out one or a small number of functions in a reliable and uniform (unvarying) manner. Even if a machine is adaptive, as in an adaptive cruise control, the adaptation must be specified in advance in order for the machine to work.

In social systems, performance variability may for practical reasons be attributed to either the individual or the social system/organisation, although organisational variability from an analytical point of view is a non-trivial result of the performance variability of individuals. It is nevertheless common practice to treat organisational variability as a phenomenon in its own right, one reason being that it in many cases would be manifestly wrong to assign the responsibility (for the possible outcomes) to an individual. (An alternative term could therefore be social variability.) Another reason is that the variability very often is due to external conditions and demands, rather than the characteristics of individuals.

In the case of humans, performance variability can occur for a number of reasons. It can be due to various physiological and psychological phenomena, such as the refractory period of cells and organs, bodily and mental fatigue, a dislike of monotony, etc. It can be due to more complex psychological phenomena such as ingenuity and self-realisation, the fact that people like to make things better, to be creative or efficient, or that they simply try to conserve resources in order to guard against undefined future developments. It can be due to socially induced variability, such as trying to meet the expectations of others, complying with informal work standards (speed, quality, etc.), trying to help – or hinder – others, etc. It can be due to organisationally induced performance variability, as in meeting demands, stretching resources, resolving ambiguity and double binds, etc. And finally it can be due to contextually induced performance variability (ambient working conditions, such as noise, humidity, vibration, temperature, etc.).

Another way to characterise performance variability is to classify it as belonging to either of the following types:

- *Teleological variability*, which occurs in situations where the goals are unstable due to external factors, such as variations in supply and demand, or where expectations or values are revised. The variability is called teleological because people adjust their performance to meet anticipated changes.
- *Contextual or situational variability*, as when people adjust what they do in order to achieve the best possible outcome – or simply a satisfactory outcome, cf. Chapter 3 – given the incomplete specification of the situation.
- Finally, a *compensatory variability*, for instance when something is missing or absent (a tool, a resource) or when the procedure for doing something cannot be remembered. (This may also be called *incidental* variability, whereas the two other forms can be called *normal* variability.)

The three types of variability correspond roughly to different levels of control. Teleological variability is in principle predictable given some knowledge of the system's environment and performance history. In teleological variability the person tries to anticipate what may happen, and it therefore corresponds to a strategic performance or level of control. Contextual or situation variability is also partly predictable, since it describes the tactics or heuristics that people use to get their work done. It therefore corresponds to a tactical performance or level of control. Compensatory variability is the least predictable, and may in fact appear as if it is more or less random, and hence corresponds to opportunistic performance or level of control. On all levels, the variability may be described using some of the ETTO rules.

Approximate Adjustments

The adjustments of tasks and activities to match actual working conditions will always be approximate and incomplete. There are two fundamental reasons for that: a lack of information and a lack of time. The lack of information is due to the simple fact that the underspecification that makes the adjustments necessary means that some information is missing. If complete information had been available when the work situation was specified or designed, then there would, in principle, be no need to adjust the work when it was carried out. Adjustments are needed precisely because information is incomplete, and this incompleteness means that the adjustments themselves must be approximate.

The lack of time happens because the work environment is dynamic rather than static. Not only is the future imperfectly known, but the time that can be spent to find out more about it is also limited. This limitation comes about both because it takes time to get information and to digest or process it, and more importantly because it is uncertain how much time is available, cf. Figure 2.2 and the discussion of 'time to think and time to do' in Chapter 2. For most people it makes sense to try to be on the safe side and leave some kind of minimal buffer of spare time. That can, however, only be done by being approximate in how the situation is assessed, hence in how adjustments are made.

Even though adjustments must be approximate, this does not mean that they necessarily will be incorrect or lead to unwanted results. On the contrary, the adjustments will in the vast majority of cases be sufficient to allow work to proceed as planned and to achieve the intended objectives. Performance variability may introduce a drift in the situation, but it is normally a drift to success, a gradual learning by people and social structures of how to handle the uncertainty, rather than a drift to failure.

Failures without Malfunctions

As discussed in Chapter 1, it has for a long time been taken for granted that adverse outcomes must be explained by failures and malfunctions of system components. All the established methods for risk assessment, such as HAZOP, FMECA, Event Trees, and Fault Trees are based on that assumption. For technological systems it is reasonable to approach safety in this way, because technological components and systems are designed to provide a narrowly defined function as reliably as possible and as long as possible, after which they fail and are replaced.

Technological components and systems function in a bimodal manner. Strictly speaking this means that for every element *e* of a system, the element being anything from a component to the system itself, the element will either function or it will not. In the latter case the element is said to have failed. This can be stated as follows:

$$e \in E, e = \begin{cases} 1\text{: component or system functions} \\ 0\text{: component or system fails} \end{cases}$$

(The criterion for whether an element functions need not be crisp but can allow, e.g., for some uncertainty or drift.) A light bulb, for instance, either works or does not work (cf. Figure 5.4). If it does not work, it is replaced since no one expects it to begin working again on its own. The principle of bimodal functioning may admittedly become blurred in the case of systems with a large number of components, and/or systems that depend on software. In such systems there may be intermittent functions, sudden freezes of performance, and/or slow drift, e.g., in measurements. However, the principle of bimodal functioning is true even in these cases, since the components are bimodal. It is just that the systems are intractable and that the ability adequately to describe and understand what is going on therefore is limited.

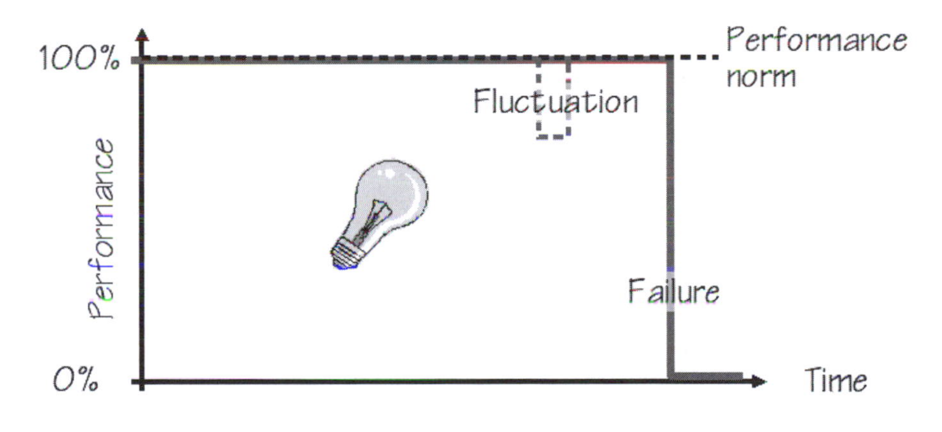

Figure 5.4: Bimodal functioning

Before the 1950s there was not a great deal of concern for human failures, or 'human error,' in the field of industrial safety. People obviously did do things wrong every now and then, but due to the nature of work they rarely harmed anyone but themselves. The extent of the unwanted outcomes was limited, and the explanations could therefore also be relatively simple, often in terms of single factors (e.g., accident proneness, lapses of attention, distraction, etc.).

This situation changed completely after the accident at the Three Mile Island nuclear power plant on 28 March 1979. The accident made the role of the human factor clear for anyone to see, and it suddenly became necessary to consider the possibility of 'human errors' in safety assessments. The simple solution was to transfer the experience from technological risk analysis to the problem of the human factor – and later also the organisational factor – leading to the need to account for 'human error,' or more precisely how human actions could be the cause of adverse outcomes. This fortuitously happened at about the time when the concept of human information–processing became *de rigueur* in behavioural science. It was therefore natural to describe the human as a machine, hence also as a machine that could fail. Indeed, the notion of a 'human error mechanism' became a favourite way of explaining human performance failures.

It is now clear that the bimodal assumption is wrong as far as humans and organisations are concerned. Humans and organisations are instead multi-modal, in the sense that their performance is variable. As seen from Figure 5.5, performance is sometimes better than the norm, sometimes worse, and on occasion even worse than the low limit, i.e., unacceptable in some way. However, performance never fails completely, i.e., a human 'component' cannot stop functioning and be replaced in the same way a technological component can. (This nevertheless does not stop people from sometimes thinking in this way.)

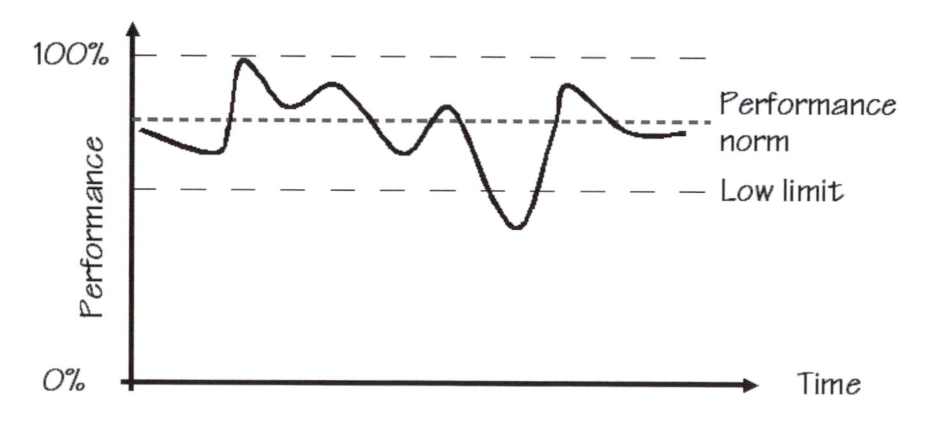

Figure 5.5: Multimodal functioning

Because the bimodal principle is wrong, it is not reasonable to explain adverse outcomes by invoking failures and malfunctions. As Ernst Mach said 'knowledge and error flow from the same mental sources, only success can tell one from the other.' This can be restated as follows: 'performance variability is the reason why things most of the time go right, as well as the reason why things sometimes go wrong.'

Another way of saying that is to point out that humans as a rule do not fail. Failures, in the sense of adverse outcomes, can happen even if nothing goes wrong. The ETTO principle is an example of that, since none of the ETTO rules are wrong as such. This also suggests that the problem of explaining how things go wrong leads us in the wrong direction. Instead we should consider how to explain how things go right, i.e., the occurrence of successes. We should try to understand and explain the normal, rather than the exceptions.

About 'Human Error Mechanisms'

The haphazard use of 'error mechanisms' as convenient ways of filling out theoretical lacunae, misses an important problem. If an 'error mechanism' is used to explain something that went wrong, the 'error mechanism' itself must either be fallible or infallible:

- If the 'error mechanism' is fallible, then the outcome will sometimes be a *correct* action rather than error, which goes against the reason for proposing the 'error mechanism' in the first place. It also means that there must be a second 'error mechanism' that affects the first, which leads to a recursive argument.
- If the 'error mechanism' is infallible, then it is necessary to invoke some further 'switching mechanism' that 'knows' when to turn the 'error mechanism' on and off, since otherwise there would only be failures. This 'switching mechanism' can itself be either fallible or infallible, which brings us back to the initial problem.

To avoid these problems it could be argued that the environment alone determines the outcome. But then there is no need to postulate the existence of an 'error mechanism' in the first place. The only sensible conclusion therefore seems to be to drop the idea of an 'error mechanism' completely.

Performance Variability and Risk

Although performance variability is both necessary and useful, there will also be situations where it can be harmful and lead to unwanted and unintended outcomes. One way of understanding that is to look at Figure 5.2 again, considering the need of performance variability on the one hand, and the risk of adverse outcomes on the other. The need of performance variability, the need of making adjustments to work – whether teleological (strategic), situational (tactical), or compensatory (operational, opportunistic) – is higher for intractable systems than for tractable systems. Since intractable systems are underspecified, the possibilities for planning and scheduling are reduced while the likelihood of interruptions or unexpected developments is increased. People at work cannot simply follow a prepared plan or set of procedures, but must constantly take the situation into account and make the necessary adjustments. The risks of adverse outcomes are furthermore higher for systems that are tightly coupled than for systems that are loosely coupled because consequences of inappropriate actions develop faster and because there are fewer opportunities for remedial actions and recovery. Together this leads to the representation shown in Figure 5.6.

Without going into every possible detail (itself an application of an ETTO rule!), the risk potential of performance variability can be characterised as follows.

- For loosely coupled, tractable systems there is little need of performance adjustments since work is fairly routine, resources adequate, and demands foreseeable. Performance variability will have limited consequences because of the loose coupling and the possibilities for recovery.
- For loosely coupled, intractable systems, work is less predictable and the need of performance adjustments is therefore higher. Due to the loose coupling, the consequences of possible inappropriate adjustments will nevertheless be limited.

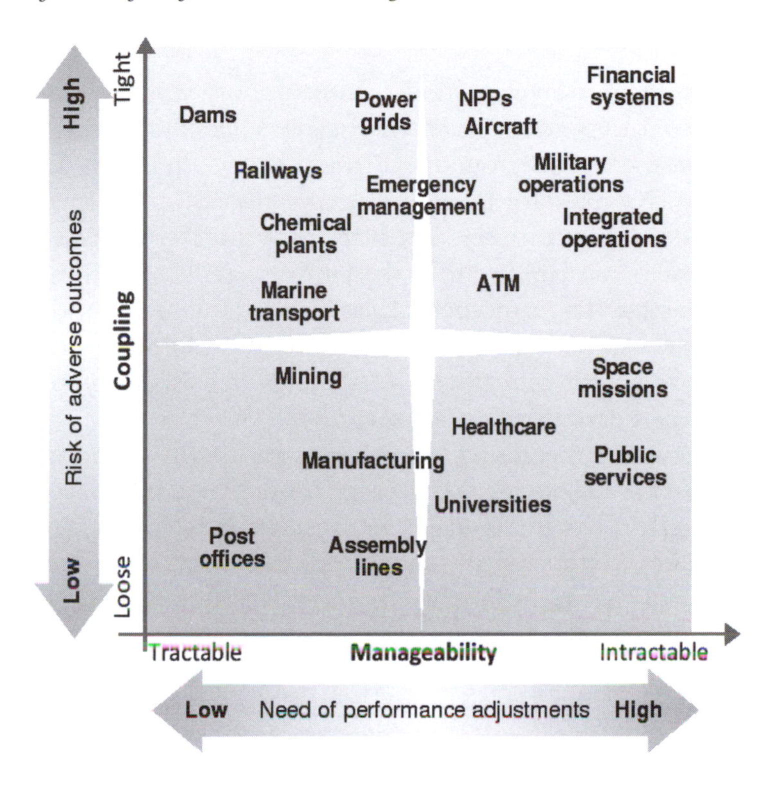

Figure 5.6: Performance variability and risk

- For tightly coupled, tractable systems, the need of performance adjustments is limited since work can in many cases be prescribed and planned. But when adjustments are made, and when performance becomes variable, the consequences may be serious because of the tight couplings. This means that effects may develop rapidly, with limited possibilities for recovery.
- Finally, for tightly coupled, intractable systems, performance adjustments are nearly always required and may in extreme cases be the only way to work. Since the systems furthermore are tightly coupled, consequences may develop rapidly and potentially be very serious with no known or available paths of recovery.

The Dark Matter

When astronomers explain their observations, for instance the rotational speed of spiral galaxies or the observed fluctuations in the cosmic microwave background radiation, they find that there is insufficient visible mass for the gravity that is needed. In order to make the theories consistent with the observations it is therefore necessary to infer that there is something else in the Universe. This something else is called dark matter. The hypothetical dark matter has much more mass than the 'visible' component of the universe and is estimated to account for as much as 96 per cent of the total energy density in the universe. Yet the nature of dark matter is unknown.

The reason for mentioning astronomy and dark matter is that it provides a useful analogy to the issue at hand. The study of accidents and 'human error' has traditionally confined itself to that which could be observed, i.e., to things that went wrong, and explanations have been developed from that basis. Yet we seem to face the same problem as the astronomers, namely that we cannot adequately explain our observations. It is, of course, possible to infer various kinds of 'human error mechanisms,' but as argued above this is not a sound solution. The problem can nevertheless easily be solved by looking at what normally is disregarded, namely what happens when things go right.

In relation to 'human error,' 'things that go right' corresponds to the astronomers' 'dark matter' but with the difference that it is something that can be observed rather than something that must be inferred. (It could therefore be called 'bright matter.') 'Things that go right' are simply the results of the variability of normal performance. To study 'things that go right,' the focus of accident investigation and risk assessment must change from looking for failures and malfunctions and instead try to understand why performance normally succeeds. This means that failures have to be seen as a special case of successes, rather than the other way around. It means that the goal of safety must change from reducing the number of adverse events to enhancing the ability to succeed under varying conditions. But it also means that there is a much broader basis to work with and to learn from.

It is not easy to estimate the proportion between things that go right and things that go wrong. One indication is that a system normally is considered as safe if the probability of an accident is 10^{-6} or lower. (In aviation safety, the odds of being killed on a single airline flight are 1 in

10.46 million or 9.5E-8. In traffic safety, the US fatality rate per 100 million vehicle miles travelled was 1.37, according to the 2007 statistics. If the average trip length is 20 miles, then the probability of a fatal accident is 2.74E-7.) This means that for every case where something goes wrong there will be about 1 million cases where something goes right. Studying why things go right will not only make it easier to understand why things go wrong, hence lead to improvements in safety, but will also contribute to improvements in efficiency, productivity, and quality.

Sources for Chapter 5

The quotation from James Reason is from his 1997 book *Managing the Risks of Organisational Accidents* (Aldershot: Ashgate).

The notion of WYLFIWYF has been used in Resilience Engineering as a perspective on accident investigation. See, e.g., E. Hollnagel, (2007), 'Investigation as an impediment to learning,' in E. Hollnagel, C. Nemeth and S. Dekker (eds), *Remaining Sensitive to the Possibility of Failure. Resilience Engineering Perspectives*, Vol. 1, (Aldershot: Ashgate). This reference also provides a good example of how an accident investigation intentionally biases the outcomes.

A reference to Charles Perrow's work on *Normal Accident Theory* was provided in Chapter 1. An important part of this is the notion of a socio-technical system. The term was already in use in the 1960s by researchers from the Tavistock Institute of Human Relations in London. The idea of a socio-technical system is that the conditions for successful organisational performance – and conversely also for unsuccessful performance – are created by the interaction between social and technical factors. (Notice the emphasis on social, rather than human factors.) This interaction comprises both linear (or trivial) 'cause and effect' relationships and 'non-linear' (or non-trivial) emergent relationships.

The usefulness of performance variability has been discussed in learning theory, for instance in relation to trial-and-error learning. A classical paper is D. T. Campbell, (1956), 'Adaptive behavior from random response,' *Behavioral Science*, 1, 105–110. Another perspective is provided by J. Reason and K. Mycielska, (1982), *Absent-minded?: The Psychology of Mental Lapses and Everyday Errors* (Englewood Cliffs, NJ: Prentice-Hall). A contemporary view is found in the topic of 'work as imagined vs. work as done,' e.g., D. D. Woods, (2006), 'How to Design a Safety Organization: Test Case for Resilience Engineering,' in E. Hollnagel, D. D. Woods and N. G. Leveson (eds), *Resilience Engineering: Concepts and precepts* (Aldershot: Ashgate).

Information about the 'real' dark matter can be found in many places on the web. It has been argued that the need to postulate the existence of 'dark matter' and 'dark energy' is an expression of human ignorance. 'Dark matter' is necessary for the current theories to explain the observations, and it is apparently simpler to suppose that 'dark matter' exists than to revise the theories. In the case of 'things that go right,' the situation is ironically almost the opposite. There is plenty of evidence to go by, but the current theories cannot account for it.

Chapter 6: ETTOs of the Past

Accident Philosophies and the ETTO Principle

Accident investigations are a clear example of the need to understand what happens around us, even if it also has created a bias of looking only at events with unwanted outcomes. Accidents are by definition events that are unexpected. That does not mean that they are unimaginable as well, although it is the case for some of them. This difference is consistent with the etymology of the term accident, which comes from the Latin *accidêns*, *ad-* + *cadere*, to fall. We all know that we can fall when we walk: it is therefore not unimaginable. But no one actually thinks it is likely to happen when it does: it is unexpected.

Since the outcomes of accidents, in addition to being unexpected also are unwanted or even harmful, it stands to reason that there is a need to understand why they happen. An 'accident philosophy' is the term commonly used to describe a way to think about accidents, especially about how accidents happen. One practically universal philosophy is based on the principle of cause–effect relations, more specifically the principle that outcomes can be described or represented as a causal chain of events that begins with a failure or a malfunction. (A causal chain can be defined as an ordered sequence of events in which any one event in the chain causes the next.) The chain of events may either be depicted as a single sequence or a path with multiple steps, or as multiple paths that interact or combine in some way.

Historically, the earliest formalised accident model is the domino model from the 1930s, which explains accidents as a linear propagation in a single chain of causes and effects (upper panel in Figure 6.1). The last domino falling is the injury, and the first is the original cause. According to this model accidents can be analysed by going backwards step-by-step from the injury, until the first cause has been found. Some 30 years later, two significant extensions appeared.

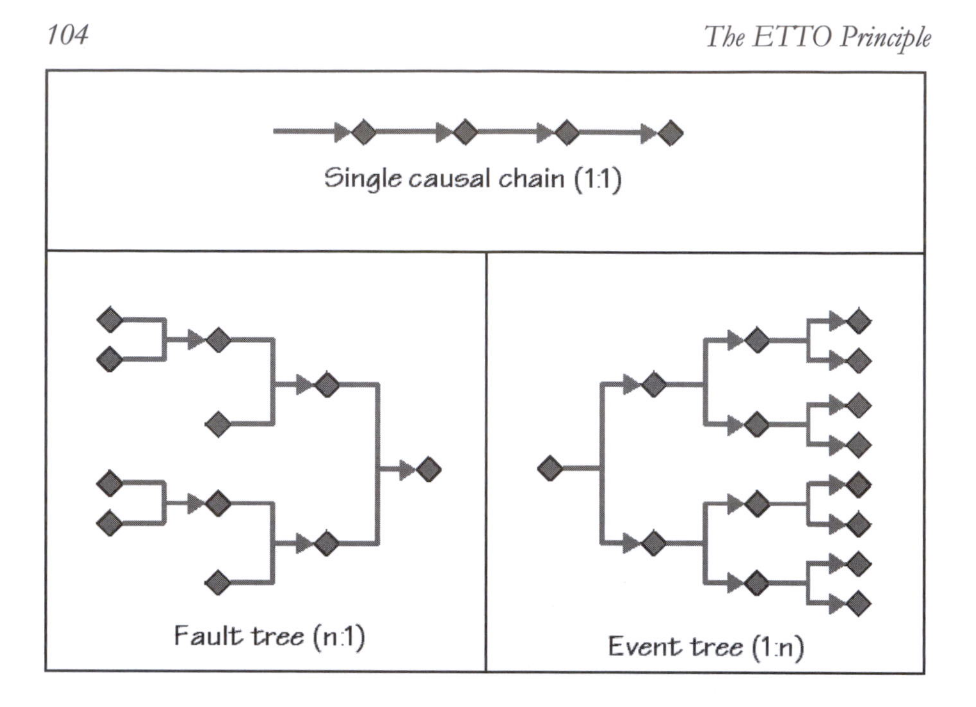

Figure 6.1: Common representations of how events develop

The first extension was the event tree, which represents accidents as a (binary branching) tree of possible outcomes from a single initiating event. (Think of it as a horizontal tree diagram with the initiating event being the single node (the 'root') to the left, lower right panel of Figure 6.1.) The direction of causality is here from the initiating event to the set of possible outcomes. The second extension was the fault tree, which represents accidents as the logical combinations of conditions and events that make up the causal chain. Fault tree analysis is essentially a deductive procedure to determine the various combinations of failures and malfunctions that can lead to the specified undesired outcomes. The direction of causality is here from the conditions and events that can lead to the final outcome (called the top event). (Think of a fault tree as a horizontal tree diagram with the top event being the single node ('outcome') to the right, lower left panel of Figure 6.1. A fault tree can graphically be a mirror image of an event tree. The two can therefore be combined, as in the so-called 'bow-tie' model.) All classical accident models make use of either single cause–effect chains (1:1 relations), event trees (1:n relations), fault trees (n:1 relations), or a combination of these (e.g., the 'bow-tie' model).

The Root Cause

Most accident models have in common that they start from the outcome or consequence, and go back step by step to the origin. There is a certain logic to doing this, since *if* an accident is the result of a linear propagation of effects in a causal chain, *then* it should be possible to reverse the propagation from the final outcome to its origin. This kind of thinking is so familiar and so common in everything we do that it is taken for granted and therefore rarely, if ever, made the subject of reflection and discussion. (To illustrate how widespread this way of thinking is, two examples will suffice. First, all religions include an account of how the world was created, usually by a single being, the creator, or from a single event. Second, and partly in contrast to that, modern physics hypothesise that the universe began with a single event, the Big Bang, and spends an inordinate amount of effort, and money, to determine exactly how this happened.)

It is a consequence of this way of thinking that it must be possible to find the first, or root, cause. A root cause can be defined as the beginning of a causal chain that ends with the specific outcome that is investigated. The attraction of a root cause is that *if* it is possible to find a single cause for any outcome, *then* the elimination or neutralisation of that single cause will prevent the outcome itself – at least according to the rationale of this approach. (Technically, the root cause must dominate over all other contributing factors. Pragmatically, the root cause defines the node in the causal chain where an intervention can reasonably be implemented to change performance and prevent an undesirable outcome.) In most cases, however, there is no single cause-event chain but rather several chains that may combine and interact, as described by a causal-tree representation. The effect of a single cause-event pair may also be contingent on the contributing factors, for instance inadequate barriers or dysfunctional defences. This corresponds to a multiple cause philosophy, where a root cause can exist for each of the contributing factors that are necessary for a resulting outcome. By the same logic, if any of those necessary causes can be eliminated or prevented, the undesired outcome will not occur.

Resultant and Emergent Effects

Root cause analysis is based on the logic and laws of causality. It is indisputable that our physical and technological world is causal or ruled by cause–effect relations, *vide* the complex machines that we are able to design, build, and operate. It is equally indisputable that we can – and do – reason in terms of cause and effect for simple and single actions or for short sequences of actions. (Indeed, the simple principle of causality is the very foundation for ETTO: if we could not reason from principles or generalised rules, we would instead have to reason from specifics on each occasion, hence we would always have to be thorough.) But we can only describe cause–effect relations for tractable systems. (Or perhaps tractable systems are defined by the fact that the cause–effect relations can be described?) It is not possible to do the same for intractable systems, which means that some other solution must be found.

Another way of characterising this difference is by distinguishing between *resultant* and *emergent* phenomena. The two terms were introduced by the English philosopher George Henry Lewes (1817–1878), who used the term 'emergent' to refer to phenomena that were new and not explicable by the properties of their components (which in modern language may be taken to mean the components of the systems that produce them). Lewes wrote that emergent effects are not additive, not predictable from knowledge of its components, and not decomposable into those components. It is therefore not possible to trace them back to a specific cause or function – or in the case of accidents, a specific malfunction or failure. (Resultant phenomena obviously represent the converse situation.)

Root cause analysis requires that all phenomena are resultant rather than emergent, since it otherwise would be impossible to carry out a causal analysis. In terms of Figure 5.6, the systems must be tractable and preferably also loosely coupled. Yet, as we have seen in Chapter 5, most systems where safety is a major concern are intractable and tightly coupled. That includes the systems where root cause analysis is mostly used, namely healthcare and hospitals. (The reader may want to refer to the case of *Incorrect Chemotherapy Treatment* in Chapter 4, to be reminded of what such systems can be like. Another example is provided later in this chapter.)

Limitations of Causal Analyses

The very fact that people tend to rely on the ETTO principle in everything they do means that causal analysis in general and root cause analysis in particular are not appropriate. People rely on ETTO rules, on approximate generalisations, precisely because it is impossible to reason or look ahead more than a few steps. Whereas classical decision theory and information–processing psychology has argued that this impossibility is due to the limitations of the human mind, the argument in this book is that the impossibility is due to the dynamic and unpredictable nature of the environment.

One part of the argument is that it is only to the extent that the last step was determined by the first step *when it was carried out*, that we can legitimately claim that the last step was caused by the first step when we look at it in retrospect. But that condition is clearly not realistic. When we look at accidents they are usually associated with long causal chains (accepting for the moment that it is sensible to talk about causal chains). We also know that for long causal chains, the outcome is unpredictable from the first step, simply because the environment is changeable rather than stable. But if the final outcome was unpredictable or indeterminate from the first step when it was carried out, then the first step – or the root cause – must also be unpredictable or indeterminate from the last step, i.e., the outcome. Otherwise it would be necessary to introduce some kind of orderliness principle that only works when events are considered in retrospect.

Another part of the argument is that the analysis is led astray by focusing on failures and malfunctions. (This is an argument against causal thinking in general, and not just against root causes.) The logic of causal analysis makes failures and malfunctions the pivotal elements, but also makes them artefacts of the causal reasoning. Apart from the logical argument, this can be easily demonstrated simply by looking at what people do in practice, cf. the many examples in this book and particularly in Chapter 4.

Causal reasoning and causal analyses are applicable only to very short cause–effect chains, e.g., a few steps at the most. For accidents that are describable in this way, the use of causal reasoning and causal analyses is justified. For other accidents, it is not. (Note the paradox that we can only know the nature of an accident after it has been analysed. However, in practice people are reasonably good at 'recognising' accident types.) Causal reasoning is in particular not justified when the circumstances are more difficult or complex.

(Relying on a simple causal analysis can, ironically, be seen as an ETTO rule applied by the analyst. While it will be more thorough to apply a non-deterministic or non-linear analysis, it will also be less efficient.)

ETTO and Learning from Experience

In order to maintain a high level of safety it is necessary to learn from experience, to learn from the results of accident investigations. Although this seemingly is a simple thing to do, it actually involves four important issues: *Which* events should be investigated and which should not? *How* should the events be described? *When* and *how* should learning take place? And *what* should the locus of learning be: individual or organisational? Here I will only look at the first issue.

Learning from experience traditionally means learning from situations or events where something has gone wrong, and often focuses on trying to find the causes of adverse outcomes, such as accidents and incidents. From a thoroughness perspective it clearly makes sense to learn from *all* accidents that happen. Although it cannot be ruled out that two accidents may have the same aetiology (cause), this can obviously not be known before they have been analysed. It would therefore be wrong *a priori* to exclude any events from the analysis. Yet because resources are limited, it makes good sense from an efficiency perspective to restrict learning to events that are significant, either in the sense that they have led to serious outcomes or because they otherwise were unusual.

Learning from experience therefore necessitates a trade-off between efficiency and thoroughness. Since no organisation has unlimited resources, it is common to limit investigations to only the serious accidents. This requires some efficient way of determining whether an accident is serious or not without having to analyse it first. An example of how this can be done is provided by the *Safety Assessment Code* (*SAC*) Matrix, developed by the US Department of Veteran Affairs as part of their *Root Cause Analysis* programme. The matrix, shown in Table 6.1, categorises accidents according to their frequency of occurrence, called probability, and the severity of the outcome. Each accident can therefore be assigned a value that is used to determine whether an investigation should take place or not. Typically, accident investigations will only consider events where the SAC value is 3.

Table 6.1: The Safety Assessment Code Matrix

		Severity			
		Catastrophic	Major	Moderate	Minor
Probability	Frequent	3	3	**2**	1
	Occasional	3	**2**	1	1
	Uncommon	3	**2**	1	1
	Remote	3	**2**	1	1

Using a rule such as this will obviously improve effectiveness because fewer resources will be spent on accident investigation. This is warranted as long as it is reasonably certain that there is nothing important to be learned from less serious accidents or incidents, i.e., outcomes with a SAC value of 2 or 1. Yet thoroughness would advocate that all registered accidents should be analysed, since it is possible that something essential may be missed by analysing only a certain subset. Returning to the *Dark Matter* argument of Chapter 5, the analysis should include performance as well. (A compromise solution from a scientific point of view would be to select cases randomly. That would increase the representativeness of events without penalising efficiency too much.) In the end, the choice of criteria may reflect political priorities and economic constraints as much as safety concerns. Regardless of how examples are selected, something will surely be learned. Yet the efficiency of learning will be decreased by reducing the thoroughness of the investigation.

Looking for ETTOs in Practice

The obvious question is *how* past events should be analysed in order to draw the right lessons from them. It is relatively easy to say that we should abandon the preoccupation with finding causes and also let go of the unspoken assumption that accidents always are due to failures and malfunctions. The ETTO principle argues that analyses should look at both what goes right and what goes wrong, because things go right and things go wrong for the same reasons. This suggests that instead of looking for things that went wrong, we should look for things that did not go right.

The practices of looking at ETTOs of the past will be illustrated by presenting another case where something went wrong in a hospital. The use of a hospital case is simply a matter of convenience. As Chapter 4 has shown, it is very easy to find cases in practically every form of human endeavour. The following case describes a situation in a hospital where a trauma patient unexpectedly arrives and must be taken care of, something that happens very regularly. Several different players, doctors and residents, are involved in receiving and taking care of the patient. In order to make the description as comprehensible as possible, the use of personal pronouns has been replaced with the full identification of the player.

1. Around 23:00 on a Friday night, *Dr A* had one hour left before her emergency room (ER) shift ended after a busy night. Her focus was to tie up loose ends in order to transfer as few patients as possible to *Dr B*, who would take over at 23:30.

2. At about that time, the trauma phone rang. A teenage girl had been hit by a car and was unconscious. As she was not breathing on her own, the paramedics were providing artificial respiration and had immobilised her neck; they should arrive in the ER in 15 minutes. *Dr A* contacted the *surgery resident* to inform him of the incoming trauma patient; she also advised the *ICU resident* that an ICU (Intensive Care Unit) bed might be required.

3. As soon as the patient arrived, *Dr A* began the systematic 'primary survey' for trauma patients. She inserted a breathing tube and verified that air was entering both lungs, she inserted an intravenous line and administered fluid, and she conducted a basic neurological examination. The examination suggested elevated pressure around the brain. *Dr A* started an appropriate treatment and ordered a head scan. She also ordered the routine trauma X-rays: cervical spine, chest and pelvis. These X-rays were done in the ER by a *radiology technician. Dr A* called the *radiology resident* and asked him to review the X-rays and the head scan. By this time, the *ICU resident*, the *surgery resident* and *Dr B* had arrived. *Dr A* explained to each the events of the accident as she knew them and what had been done so far. She told them that the scan had been ordered and that the *radiology resident* would review the films. She transferred the rest of her patients to *Dr B*, finished her paperwork, and went home. The rest of the examination and management would be taken care of by *Dr B* and the residents.

4. *Dr B* had a quick look at the patient and determined that she was stable. He suggested that the *ICU resident* should accompany the patient to the scan and after that directly continue to the ICU in order to free a place in the ER. Satisfied that subsequent care would be assured by the *surgery resident* and *ICU resident*, *Dr B* signed off the case and started seeing new patients. The *radiology technician* called the ER to let *Dr A* know the films had been processed, but by that time *Dr A* had already left.

5. When *Dr A* called the *surgery resident* (step 2), the latter was having a quick dinner after three emergency operations. He had yet to round on the in-patients on the surgical ward. Since *Dr A* had examined and stabilised the patient, the *surgery resident* did not see a need to repeat what had already been done. Though the surgery team was officially 'in charge' of all trauma patients for 24 hours, there was less for the *surgery resident* to do in the case of an isolated head injury. It would suffice to check that there were no other injuries and coordinate care with the ICU and the *neurosurgeon*. On his way down to the ER, the *surgery resident* met the *ICU resident* and together they took the handover from *Dr A*. The *surgery resident* knew the *ICU resident* would take care of adjusting the ventilator and managing medical treatments for the head injury. They agreed to meet in the head scan room and the *surgery resident* ran up to the ward to check on a few patients.

6. When the *surgery resident* returned to the scan room, the scan had already been done. The scan did not show any bleeding or fracture but there was a degree of swelling. This could be managed by the *ICU resident* together with the *neurosurgeon*. The chest, spine and hip X-rays were up on the screen. The *surgery resident* looked at them quickly. They seemed OK, the vulnerable part of the spine was seen on the CT scan, there was no reason to suspect any other problems, and the *radiology resident* had seem them. The *surgery resident* looked at the lab results but there were no suggestions of internal bleeding or injury to other organs. The *surgery resident* confirmed that this patient had an isolated head injury and that the patient from then on would be managed by the *ICU resident* and the *neurosurgeon*.

7. The *ICU resident* had just finished admitting another new patient to the ICU when he got the call from *Dr A* (step 2). He thought about the important points of managing a head injured patient and wondered if an intracranial pressure monitor would be required. To do so was an invasive procedure and the *ICU resident* would have to present his

reasoning clearly to the *neurosurgeon*. At least *Dr A* had already put in the breathing tube, which last time had been a difficult procedure. The *ICU resident* let the nurses know the patient was coming, what equipment and medications to set up, and joined the *surgery resident* on his way down to the ER (as described in step 5). The *ICU resident* took in the information from *Dr A* and assessed the patient. The ventilator seemed to be working fine, the chest X-ray was normal, the breathing tube was in good position, and the patient's oxygen level was good. The patient's blood pressure was normal and though the patient had been sedated, there were no longer clinical signs of increased pressure around the brain. The *ICU resident* quickly examined the rest of the patient, but did not find any bruises, any deformities, or any masses. The *ICU resident* assumed that the *surgery resident* would take care of the detailed head-to-toe exam later. The *ICU resident* had to go to see another patient and the scan room was calling for the patient.

8. The head scan showed no bleeding or fractures. There was a mild degree of swelling and the *ICU resident* was unsure if an internal pressure monitor would be necessary. As he was waiting to speak to the *neurosurgeon*, the *ICU resident* took a quick glance at the pelvic X-ray that was up on the screen. It looked OK, but the *surgery resident* and *radiology resident* were better at interpreting pelvic films.

9. After a lengthy telephone discussion with the *neurosurgeon*, it was decided that a pressure monitor was not necessary. A management plan was reached and the *ICU resident* brought the patient up to the ICU. The *ICU resident* reviewed all the lab tests, set up the ventilator and put in another IV line, ordered IV fluid, antibiotics, and painkillers, and called his *attending physician* to inform him of the admission. The plan was not to oversedate the patient so that she could wake up if possible; this would permit a better evaluation of her neurological state. The *ICU resident* wrote the medical orders, spoke with the nurses and the patient's parents. The *ICU resident* then went about seeing the rest of the patients in the ICU.

10. The *radiology resident* answered his page from *Dr A* and came back to the hospital to interpret the head scan. The *radiology resident* spent quite a bit of time examining the scan and discussing the results with the *ICU resident*. Because the *radiology resident* was there, he had a quick look at the other X-rays but did not spend too much time on it – the other residents had already looked at them, and the only reason he had to be in the hospital was for the scan. Normally a *radiologist* would

not look at plain X-rays on a Friday night; it could wait until the next day.

11. The next day was a Saturday. The *ICU resident* transferred the ICU patients to the new team and went home for a well-deserved rest. The *surgery resident* stopped by and saw that everything seemed to be under control. The *neurosurgeon* performed a limited exam, reviewed the head scan, and agreed with the ICU management. The fresh ICU team saw the patient on their rounds that morning and reviewed the head scan with the *neurosurgeon*. The patient was starting to wake up but was still a bit too drowsy to remove the breathing tube. Her vital signs suggested she was in pain and her pain medication was adjusted accordingly.

12. The *radiology attending* arrived Saturday morning and began reviewing the films from the night before. He noted that the *radiology resident* must have reviewed some films that still were up on the screen. The *radiology attending* dictated his reports into the hospital system – the dictations could then later be accessed by the treating teams.

13. On Sunday, the patient was breathing well on her own and the breathing tube was removed. She complained of a headache and aches 'everywhere.' Her neurological examination was normal and her pain medication was adjusted.

14. On Monday, the *radiology technician* who had performed the X-rays in the ER bumped into the *orthopaedics resident* in the cafeteria. The *radiology technician* remembered the dislocated hip she had noticed on a film she took on Friday. She wondered why she had not been called back to take another film after the hip had been put back into position and asked the *orthopaedics resident* about it. The latter was unaware of any patient with a dislocated hip. It was subsequently discovered that the pelvic X-ray of the head-injured patient did show a dislocated hip; although it had been missed at the time, it was very obvious in retrospect. Because the hip had not been put back into position quickly, the blood supply to the joint had been cut off. The teenage girl recovered from her head injury, but she would need an artificial hip.

Even a quick reading of this case shows several instances where the people involved traded off thoroughness for efficiency. By now, this is hardly a surprise to the reader. A more conventional analysis of the case would focus on each of these instances and try to explain how the person, or persons, in question had failed, either by invoking a number of 'error mechanisms,' by naming various types of violations, or by looking for contributing factors and their root causes. But as argued above, we should not try to understand the events of the past by focusing on what went wrong. Instead we should try to understand what should have happened and how work normally should have taken place. We should then look for the trade-offs and adjustments that are normally made, i.e., the reasons why things usually go right. From that basis it becomes possible to understand why things in this specific case went wrong. In other words, the focus should not be on failures but on what should have gone right, but which did not.

Collaborative ETTO

There is a significant difference between things that went wrong and things that did not go right. The explanation of an unintended outcome ('something gone wrong') can, of course, be found in the incorrect execution of a single action ('failure'), but is more likely to be found in the interaction among performance variabilities. This underlines that actions and activities, such as the ones described in the case above, do not happen in a social vacuum, but rather are part of a complex fabric. What one person does follows what another has done, and is itself followed by what a third person will do (Figure 6.2). These couplings may be more or less immediate, as in the current case, or happen with long delays – often so long that the couplings are missed (cf. the loss of M/V *Estonia*, for example).

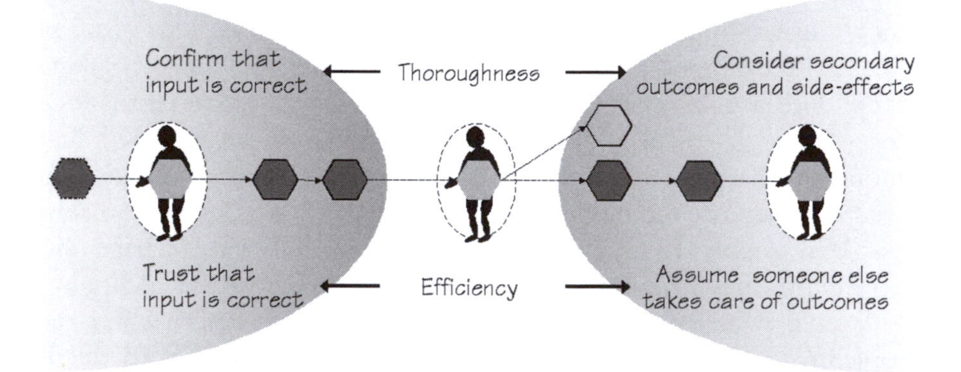

Figure 6.2: ETTOing in a collaborative setting

Any person may be a part of this social fabric, or distributed work. In such cases, thoroughness means that the person does not simply accept the input he or she receives from somewhere or from someone else, but instead makes an effort to confirm that it is correct. (This should also apply even if the input comes from a trusted source, cf. the *Publishing of Aeronautical Data* example.) Thoroughness would also mean that people consider the possible side-effects and secondary outcomes of what they produce as output, in a sense adopting the mindset of whoever is going to work on the results. Similarly, efficiency means that people trust that the input they receive is correct, i.e., that the previous person was thorough. Efficiency also means that the person assumes that the next person, whoever is going to work on the results, will make the necessary checks and verifications, i.e., that the next person is thorough. This will lead to a situation as shown in Figure 6.3.

Figure 6.3: Mutual assumptions about ETTOing

For distributed work, for the social fabric, there is never enough time to check everything. It is therefore necessary to trust what others do. Trading off thoroughness against efficiency is in practice the only means available to save time and resources. In the best case people may think ahead – or back – one step, but only spend a small amount of effort to do more than that. It is as if everyone, including ourselves, reasons in the following way: 'I can allow myself to be efficient, because the others will be thorough.' If only some people do that, the system may be able to correct itself and to find a balance of functioning that is both reasonably effective and reasonably thorough. But if everyone begins to work in this way, for instance because of systemic pressures, the net result may be that something goes wrong. Yet the reason for that is not that anyone did anything that was manifestly wrong or incorrect. The reason is more likely that everyone did their own bit of ETTOing, quite as they normally do.

What Really Happened?

With these principles in mind, we will try to understand what happened by focusing on what did not go right. While the events can be categorised in several ways, for the sake of the discussion we will use the following: alert (steps 1–2), admission and primary survey (steps 3–4), detailed survey (steps 5–8), admission to the ICU (step 9), follow-up survey (steps 10–13), and recovery (step 14). It is obvious, even to a non-professional, that each step normally involves a number of players, and that the whole procedure is complex in any meaning of the word. It is also clear from the description that the work is subject to a number of external conditions and pressures, which themselves result from trade-offs made on the level of the organisation.

Although it is not stated clearly in the description of this case, it does not seem unreasonable to assume that hospital management supported a policy of 'safety first.' This clashes with the pressure to keep waiting room times to a minimum (step 1), to clear out the ER quickly (step 4), and to avoid unnecessary checks (step 11). There also seemed to be a shortage of residents, most of whom were overworked and overtired.

The alert seemed to function as it should, but from thereon a number of things did not go as well as they should have. The reason was not that the people involved made 'errors' or mistakes, but rather that they for various reasons were unable to be sufficiently thorough. The admission and the primary survey were carried out under time pressure. In the case of Dr A because she was ending her shift, and in the case of Dr B because he had to attend to other patients. (New admissions to the ER would presumably be arriving more or less continuously.) Dr A initiated a number of actions, but did not follow up on them ('*someone else will do it later*'). Dr B only briefly looked at the patient ('*it looks fine*'), delegated further work to others ('*someone else will do it later*'), and then attended to his other duties.

The detailed survey was also subject to multiple trade-offs. The surgery resident was obviously overworked and therefore tried to avoid doing something that others had already done before (step 5), or that others would do later (step 6). The same was the case for the ICU resident, who checked what Dr A had done but otherwise accepted it, and who only made a quick examination of the patient since the surgery resident would take care of that later (step 7).

The admission to the ICU, seems to have worked as intended, but the follow-up survey did not. By that time everyone, quite rightly, worked on the assumption that the head injury was the critical issue. The prominence of the head injury nevertheless distracted from a thorough systematic evaluation of other potential injuries:

> 'This represents a confirmation bias, a strong psychological phenomenon that is found everywhere. There is hardly any situation where it is not easier to look for confirming evidence ("*it looks like X, so it probably is X*") than to look for falsifying or contradictory evidence. The former allows you directly to look for what you already have in mind, while the latter requires several steps of reasoning and reflection before a search for falsifying evidence can begin. In the social situation at the hospital each player also assumed that others had been thorough in their examination, hence that they could save some time and effort. They could justifiably do so, since they themselves were thorough, or as thorough as they thought necessary, in their own examination.'

The follow-up survey also took place over several days, hence was interspersed with other activities; this means that it was more difficult to re-examine a fact than if everything had taken place in a single (synchronous) setting.

The recovery, finally, was fortuitous rather than deliberate. In this case the radiology technician was concerned that the work she was responsible for had not been completed (step 14). The concern was, however, not strong enough to warrant a direct intervention in the system, but was only expressed when the opportunity arose ('*it is not really important*'). The same mindset seems to have been present when the films had been processed, but Dr A already had left (step 4). The handover between different players, through which the responsibility for the patient was transferred from one organisational unit to another, made use of opportunities (people meeting each other) rather than a formal procedure. This was probably more efficient, but may also have been less thorough, in the sense that some things were simply taken for granted.

Small and Large Causes

The description of the case given above was produced following certain principles, although the account in itself is not intended to be a description of an analysis or investigation method. The description has also deliberately refrained from producing a graphical representation of the events, although this is quite possible. Readers who are interested in how this may be done will find some suggestions in the *Sources* section of this chapter.

In retrospect, the critical weakness of how the patient was treated, was that the hip dislocation was missed in the pelvic X-ray, despite the fact that the X-ray was examined several times. In each instance one or more ETTO rules can be found at play. (The reader is invited to determine which.) Both Dr A and Dr B. failed to check the pelvic X-ray. Neither the surgery resident nor the ICU resident nor the radiology resident noticed the hip dislocation on the X-ray, even though they looked at it. The ICU team that started work on Saturday morning failed to check the X-ray. The radiology technician did not inform anyone of the hip dislocation, even though she had noticed it. And finally, although the radiology attending saw the dislocation, he just dictated the report but did not contact the treating physicians.

This case illustrates something very typical, namely that there are many 'small causes,' but few 'large causes.' The 'small causes' are furthermore something that neither can nor should be addressed one-by-one or in isolation. It may, of course, be tempting to blame each individual for their 'failures,' and in doing so hope that something like this will not happen again in the future. Yet by assigning blame or pointing to specific causes we disregard the fact that the whole system only worked because everyone made approximate adjustments to their work. The problem for the analysis is to understand how these adjustments affected each other and affected the development of the event, rather than to understand any adjustment by itself. The real power of an analysis such as the above is the understanding that what *anyone* does depends on what others have done, what others do, and what others are going to do – and that *everyone* is in the same situation. In order to be able to improve safety we must acknowledge that the most important part of the – dynamic and unpredictable – environment is what other people do. It is the ability of people to adjust their performance in response to the adjustments that others have made, make, or are going to make, that makes the social system strong. But it is also that which makes it weak. The performance of the system as a whole is surprisingly stable and effective because there are continuous adjustments on all levels, but sometimes also fragile because the adjustments are approximate. Learning from the ETTOs of the past must therefore focus on what makes the system strong, not on what makes it weak. It is only in this way that we can ever hope to improve safety.

Sources for Chapter 6

The domino model dates from Heinrich's book on *Industrial Accident Prevention* from 1931, as already mentioned in the sources for Chapter 1. A general introduction to the history of accident models can be found in E. Hollnagel, (2004), *Barriers and Accident Prevention* (Aldershot: Ashgate).

Event trees and fault trees both date from the early 1960s. The first known application of fault trees was to evaluate the Minuteman Launch Control System for the possibility of an unauthorised missile launch. The combination of fault trees and event trees was described already in the 1970s using the name of cause-consequence trees, cf. J. R. Taylor, (1976), *Interlock Design Using Fault Tree and Cause Consequence Analysis* (Risø National Laboratories, Risø-M-1890/R-7-76). The same principle was used as the basis for what is commonly known as the 'bow-tie' model, which has been described by, e.g., J. P. Visser, (1998), 'Developments in HSE management in oil and gas exploration and production,' in A. R. Hale and M. Baram (eds), *Safety Management: The Challenge of Organisational Change* (Oxford: Pergamon).

Root cause analysis is both a widely used general term and a specific accident analysis method. Details about the method can be found at http://www.patientsafety.gov/rca.html. Root cause analysis is widely used and is supported by extensive instructional materials and practical tools. It should, however, only be used when the conditions allow, i.e., for tractable and loosely coupled systems.

I am also grateful for the permission from Dr Karen Harrington, MDCM, from the CHU Sainte-Justine Université de Montréal, Montreal, Canada, to use excerpts from her analysis of the head injury/hip dislocation case. Any mistakes or misinterpretations in the presentation of this case are obviously mine.

As far as a more systematic approach to analyse such events is concerned, one candidate is the Functional Resonance Analysis Method (FRAM). A general description of the method is given in E. Hollnagel, (2004), *Barriers and Accident Prevention* (Aldershot: Ashgate). Examples of how this method has been used can be found in e.g., T. Sawaragi, Y. Horiguchi and A. Hina, (2006), *Safety Analysis of Systemic Accidents Triggered by Performance Deviation* (SICE-ICASE International Joint Conference, Bexco, Busan, South Korea), or D. Nouvel, S. Travadel and E. Hollnagel, (2007), *Introduction of the Concept of Functional Resonance in the Analysis of a Near-accident in Aviation* (33rd ESReDA Seminar: Future challenges of accident investigation, Ispra, Italy).

Chapter 7: ETTOs of the Future

I believe that it is important that we recognize that although it is impossible to predict the future, the one thing that is certain is the uncertainty of it.
Tony Blair, Parliamentary debate, 14 March 2007 (on whether the UK should maintain a nuclear deterrent.)

As discussed in Chapter 1, it is a truism that everything that can go wrong will go wrong. However, as a truism it is a cliché rather than a self-evident truth. Indeed, if experience is anything to go by then most things that can go wrong will usually go right. Accidents, and even incidents, are on the whole very rare except for high-risk activities such as commercial fishing and open heart surgery, and even here the risk is still around 10^{-3}, i.e., one time out of a thousand. Accidents in safe and ultra-safe systems happen so rarely that most people go through life without ever experiencing one themselves. (In road traffic, for instance, the US fatality rate in 2007 was 1.37 per 100 million vehicle miles travelled.) The worry is therefore not that things go wrong very frequently, but rather that the consequences when it happens can be so serious that it is worth the effort to try to find out in advance. This is a legitimate concern both in safety critical industries, in finance and commerce, in private and public transportation, in manufacturing and production (where it often is expressed in terms of quality assurance and quality management), in healthcare, etc.

In order to find out what can go wrong it is necessary to be able both to understand the events of the past and to predict what may happen in the future. The former has been discussed in Chapter 6, and the latter is the subject for this chapter. Everyone realises that the future is uncertain, as the above quote illustrates, although we generally need to do more than just acknowledge the fact. In order to be able to eliminate or prevent risks, or to be adequately prepared for the unexpected, we must be able to predict what may happen in the future with a reasonable degree of confidence.

Classical risk assessment, as described in Chapter 5, is concerned with determining either how likely it is that a component or a subsystem may fail or malfunction, or how a specific unwanted outcome may possibly come about. For socio-technical systems these are however the wrong questions to ask. A more appropriate question is how and when the variability of normal performance, i.e., the adjustments that people must make to accomplish their work, can lead to adverse outcomes. Just as Chapter 6 illustrated how we can understand the events and ETTOs of the past, so will this chapter illustrate how we can begin to understand and possibly predict the ETTOs of the future. The following sections will present the principles for how this can be done, rather than a ready-made method. The reason for that is that we do not primarily need new methods, but rather new ways of thinking about why things mostly go right but sometimes go wrong. Existing methods should be used as far as possible, but in a way that is consistent with the changed perspective.

This chapter will more concretely consider two issues. The first is how the ETTO principle can be used to provide a better understanding of risks, hence in some small way contribute to reduce the uncertainty of the future. The second is that every method represents an efficiency-thoroughness trade-off and that this constitutes a setting for the method, and hence serves as a basis for using it more appropriately.

The Framing of the Future

Predictions of what can possibly go wrong are traditionally linked to the notions of failures and malfunctions. One part of risk assessment, or risk analysis, is to determine what can go wrong in a system, or more specifically to identify the combinations of initiating events, internal conditions, and external conditions that may lead to adverse consequences. A second part is to characterise the possible adverse consequences and estimate how severe they can be. A third part is to calculate how probable these adverse consequences are, typically by charting the possible ways in which failures and malfunctions can combine and using this to structure the calculations. Risk assessment can in principle start either from the failure of a component or a subsystem and go on to determine the possible outcomes, or from a specific undesired outcome and determine how this could possibly come about. The former is exemplified by an event tree and the latter by a fault tree, cf. Figure 6.1.

Since we customarily think of past events as being either correct or incorrect (the bimodal principle, cf. Chapter 5), it is hardly surprising that we tend to think of future events in the same way. As far as future outcomes are concerned it is, of course, reasonable when they occur to apply some kind of distinction between what is acceptable and what is unacceptable, although the threshold or criterion may vary from time to time or between people, and depend heavily on the current conditions. (To illustrate that, consider how an accident may be defined in different domains or by the same domain in different cultures. In the Nordic countries, for instance, a train accident is defined as a situation where railway equipment has been in movement and where people were killed or seriously injured or the damage value is over €10,000. In Japan, a large train accident is defined as one that either involves a loss of more than ¥500,000, about €4,000, or causes a delay of more than 10 minutes to the first bullet train of the day.) The criterion may also change as events develop because time and resources are limited (cf. the *satisficing* principle described in Chapter 3), and may even be changed after the outcome has occurred in order to justify the actual outcome (cf. the mentioning of *post-decision consolidation* in Chapter 2).

The bimodal principle is indirectly reinforced by the concept of ultra-safe systems. This term refers to systems where the safety record reaches the mythical threshold of one disastrous accident per 1 or 10 million events (10^{-6} or 10^{-7}). Such systems are called ultra-safe because the probability of a failure is very low, but not because of the way in which this is achieved, for instance because they are High Reliability Organisations (HRO) or resilient systems. The focus is, as always, on the product (outcome) rather than on the process. Yet it would be more in line with the principles of Resilience Engineering if ultra-safe systems were named because their ability to recover from adverse events was ultra-high, rather than because the probability of failure was ultra-low.

While it makes sense from an efficiency point of view to treat outcomes as either acceptable or unacceptable, the same cannot be said for the events that may lead to the outcomes or be judged to be their causes. First, to repeat the paraphrase of Ernst Mach's famous dictum, *success and failure have the same origins, only the outcome can distinguish one from the other.* This means that it is unreasonable to assume that different outcomes are due to different types of processes, hence to different types of events. (This is also in good agreement with the Law of Parsimony, which states that the explanation of any phenomenon should make as few assumptions as possible.) Second, as the ETTO principle argues, human performance cannot reasonably be categorised as either right or wrong, for instance as a normal action or a 'human error.' Human performance is always variable and always adjusted to the conditions of work. It may well be common practice to rely on a binary categorisation when explanations are sought, but that does not mean that it has any scientific value as part of a method or theory.

In consequence of this it is necessary to abandon the tradition of basing risk assessments on the notion of failures, at least when the assessment includes human and organisational functions. Instead we must acknowledge that there can be situations where failures as such cannot be found, but where the outcome nevertheless is unacceptable. It is also necessary to refrain from relying on the principle of linear combinations of effects. This follows from a simple apagogical (*reductio ad absurdum*) argument: if adverse outcomes were due to a linear combination of effects, then there would have to be a failure or a malfunction for the chain of events to start. But since the latter is no longer a necessary condition (given that the ETTO principle is correct, of course), then there is no need to assume that effects combine linearly. (This argument applies regardless of whether the causal chain depicts 1:1, 1:n, n:1, or n:n relations.)

Understanding the Risks of the Future

The basic question in most of the established risk assessment methods, is how likely it is that something will malfunction or fail. This question may be meaningful for a technological system, but becomes problematic when the assessment addresses human actions, i.e., when it is of a socio-technical system. For such systems the question of a component failure does not make much sense, because the 'components' – humans and organisations – do not fail in the conventional meaning of the term. Unfortunately, all systems of interest today are socio-technical systems, which means that they depend on the effective interaction and collaboration between humans, technologies, and organisations.

Instead of asking how likely it is that a person will do something wrong, the question could be how likely it is that a person or an organisation will make an efficiency-thoroughness trade-off. The answer to that is that it is not only very likely but almost certain, because making these trade-offs or adjustments is the very basis for work. Not only will people always *be* ETTOing, but they *should* do so.

This answer therefore suggests that the initial formulation of the question was wrong. The question should instead be this: under which conditions is it likely that the aggregated or combined effects of ETTOing result in unintended, adverse outcomes? Or in other words, when is the mutual and interdependent system of individual and organisational performance adjustments likely to produce an unwanted outcome?

ETTO Rules as Situational Response

The new formulation of the question changes the problem from a search for situations where people will rely on an ETTO rule to whether it is possible to anticipate how ETTO rules are affected by the conditions. In other words, which kinds of efficiency-thoroughness trade-offs are people likely to use in which situations? Since we already have a list of possible ETTO rules to start from (Chapter 2), the question is how different situations can best be characterised. The ETTO rules, as they have been formulated, serve the purpose of efficiency, which means that a natural first step is to consider situations where the resources needed to meet the demands of the situation are limited. One important resource is clearly time. An indication of how this may affect performance can be obtained from the study of reactions to information input overload mentioned in Chapter 3. More generally, a lack of resources can lead to an inability to do something in time, the resources being knowledge, experience, tools, materials, etc. It is also necessary to consider what the possible consequences of using an ETTO rule are, for instance in terms of how thoroughly the preconditions for doing something are checked or how effectively the execution of an activity is controlled – including a consideration of the possible side-effects. Table 7.1 illustrates how such an understanding can be expressed, but the relations shown should be seen as tentative rather than definitive.

One ETTO rule that does not quite fit into Table 7.1 is 'if you don't say anything, I won't either.' In this situation one person has 'bent the rules' in order to make life easier for someone else. The 'bending' itself may match one of the rules in Table 7.1, but this rule is special because it is social rather than individual. (There may, of course, be other examples of social ETTO rules.)

Table 7.1: Condition specific ETTO rules

A shortage of ...		... may make a person rely on an ETTO rule, such as ...	... which means that ...	
time	resources (in particular information)		precondition checks may be	control of execution may be
Short	Inadequate	Looks fine / it is not really important / we always do it in this way here	brief	limited
Short		Not really necessary	brief	
Short		Normally OK, no need to check	brief	
Short		Will be checked by someone else		limited
Short		Has been checked by someone else	brief	
Short	Inadequate	No time, no resources – do it later / we must not use too much of X		limited
Short		We must get this done / it must be ready in time / ... this way is much quicker	brief or absent	limited
Short		It looks like Y, so it probably is Y	brief or absent	limited
	Inadequate	Can't remember how to do it		limited
	Inadequate	It worked last time	brief	
Short	Inadequate	This is good enough for now		limited

This overview of how ETTO rules correspond to situation characteristics must be developed further in order to provide a practical basis for socio-technical risk assessment. This can be done, e.g., as a field study, by going through existing 'raw' event reports, or simply by using the experience of seasoned investigators. An additional step should be to characterise the situation or context in more detail than done here, i.e., to go beyond the two obvious factors of 'time' and 'resources.' It seems reasonable to assume that social factors such as 'trust,' 'organisational climate,' and 'social norms' also play a role, in addition to more traditional factors such as 'training,' 'ambient working conditions,' 'adequacy of equipment and interface,' etc. Because the efficiency-thoroughness trade-off is a general human trait rather than a response to a unique work situation, most of the factors will be relatively domain or application independent. It is, however, conceivable that there also may be some context-specific factors due to the nature and environment of specific types of work.

ETTO at the Sharp End and at the Blunt End

In risk assessment, as well as in accident investigation, the focus is usually on the people at the sharp end, i.e., the people who are working at the time and in the place where the accident happens. It is, of course, recognised that there also are people at the blunt end, meaning those who through their actions at an earlier time and in a different place affected the constraints and resources for the people at the sharp end. (The distinction between 'sharp' and 'blunt' is clearly relative rather than absolute.) But models and methods available for the analysis of blunt end activities are but poor cousins of those available for the analysis of sharp end activities.

This difference is undoubtedly an artefact of the ways in which both accident analysis and risk assessment are carried out, the first starting with the accident that happened and the latter focusing on the possible unwanted outcomes. Classification schemes for 'human error' are almost exclusively directed at actions at the sharp end, i.e., at 'operators' rather than 'managers.' Yet from an ETTO point of view, people obviously make the same kinds of trade-offs in their work regardless of where they are. How they try to make ends meet may therefore differ less than the conditions that affect their performance. In both cases a shortage of time is clearly an important factor, cf. Table 7.1. In addition, work at the blunt end often suffers from a lack of information, simply because managers often are removed – in time and in space – from the actual operations. At the sharp end, activities usually have a clear beginning and end, information is concentrated, and the work environment includes specialised support (tools, training, etc.). At the blunt end, work is often scattered, information is dispersed, and the work environment is generic with little specialised support. Indeed, one might say that the system at the blunt end is less tractable than at the sharp end, and that ETTOing therefore plays a larger role. Since risk assessment should consider performance variability as well as performance failures, the efficiency-thoroughness trade-offs at the blunt end are at least as important as those at the sharp end.

The Nature of Emergent Outcomes

Understanding – and anticipating – which ETTO rules people may use in specific situations is only one part of the answer. The other part is to understand how the variability of everyday performance may sometimes combine and lead to outcomes that are both unexpected and out of proportion to the individual performance adjustments. We have many methods to predict point failures and account for linear combinations. These methods have the following principles in common:

- Systems can be decomposed into meaningful elements (typically components or events).
- The elements are subject to the bimodality principle, and the failure probability of elements can be analysed/described individually.
- The order or sequence of events is predetermined and fixed, usually given by how the system is organised. This assumption is clearly illustrated by event and fault trees, which depict cause–effect chains.
- Combinations of outcomes can be described as linear, i.e., they are proportional to the causes and predictable from the causes.

Risk assessment traditionally relies on representations of events that have been explicitly defined by the analysts, such as trees, hierarchies, and nets, and assumes that developments will take place as described by the representation. However, these methods cannot be used when nothing as such goes wrong and when the systems we consider are intractable. The risks of socio-technical systems can neither be analysed nor managed by considering only the system components and their failure probabilities. In order to think about what can happen in intractable systems there is a need of more powerful methods, and in particular of methods that offer an alternative to cause–effect thinking. For such systems, the premises for risk assessment must look something like the following:

- Systems cannot be described adequately in terms of their components or structures.
- Neither the system as a whole, nor the individual functions can be described as bimodal.
- While some outcomes may be determined by the failure probability of components, others are determined by interactions among the variability of normal performance.
- Risk assessment should try to understand the nature of the variability of normal performance and use that to identify conditions that may lead to both positive and adverse outcomes.

Since performance variability or ETTOing is an everyday phenomenon, it will rarely itself be the cause of an accident or even constitute a malfunction. But the variability of multiple functions may combine in unexpected ways that lead to disproportionally large consequences, hence produce non-linear effects. Such outcomes are called emergent, as explained in Chapter 6. (The outcomes may, of course, be either positive or negative.) The strength of socio-technical systems is that they can change and develop in response to conditions and demands. But this also makes them intractable in the sense that it is impossible to describe all the interactions in the system, hence impossible to anticipate more than the most regular events. The interactions are mostly useful, but can also constitute a risk.

The Small World Problem

One way of understanding how emergent outcomes can come about is to consider the so-called 'small world' problem (Figure 7.1). This is usually formulated as follows: How likely is it that any two persons, selected arbitrarily from a large population, can be linked via common acquaintances and how long will the links be on average? (In risk assessment, the problem is how likely it is that two events are indirectly coupled and how many steps in between are required on average.)

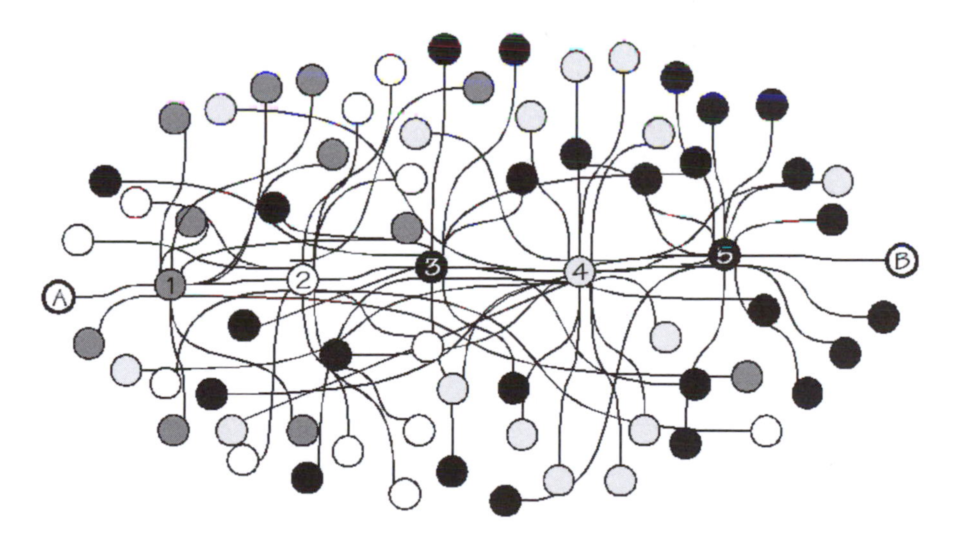

Figure 7.1: Six degrees of separation

The length of the links, or the number of steps between two persons A and B, is also called the degree of separation. It refers to the idea that a person is one 'step' away from each person he or she knows, two 'steps' away from each person who is known by one of the people he or she knows, 'three' steps away from each person who is known by one of the people who is known by a person he or she knows, etc. By repeating this argument it becomes clear that no one is more than six 'steps' away from each person on Earth. (The mathematics of that are simple: if everyone knows at least 45 others, then only six steps are needed, since $45^6 = 8.3$ billion, which is roughly the number of people on the planet.)

The point of the argument is that since we do not know precisely how the parts of a socio-technical system relate to one another, and in particular do not know precisely how activities relate to one another, then we should not be surprised if two seemingly unrelated persons or activities in fact may be coupled to each other. Technical systems consist of components with designed links and relations that do not normally change. But socio-technical systems are not designed and may hide uncharted and unknown dependencies or couplings. The small world phenomenon demonstrates the importance of this, namely that things (actions) that seemingly have no relation to each other still may affect each other. Such effects are of course made stronger if couplings can interact with and reinforce each other. The critical issue in risk assessment it not just that the number of combinations is too large to handle. The critical issue is that we need to be able to account for the couplings that emerge in socio-technical systems in order to be able to understand the risks of the future.

Risk Assessment as Efficiency-Thoroughness Trade-Off

Probabilistic Risk Assessment (PRA) is a widely used methodology for risk assessment in complex engineered systems. (Since the late 1980s it has also been known as Probabilistic Safety Assessment or PSA.) In the late 1940s and early 1950s, advances in electronics and control systems led to a rapid growth in the complexity of many activities, both military and civil. This made it necessary to ensure that the new systems were able to function as intended, using probability theory as a basis for the new field of reliability engineering.

PRA was soon used for nuclear power plants (NPP) to develop scenarios for hypothetical accidents that might result in severe core damage, and to estimate the frequency of such accidents. The landmark study was the WASH-1400 report from 1975, which used a fault tree/event tree approach to consider the course of events which might arise during a serious accident at a large modern Light Water Reactor. Since then PRA has become the standard method for safety assessments in NPPs, covering all phases of the life cycle from concept definition and pre-design through commercial operation to decommissioning, and has also been taken into use in other safety critical domains.

The efficiency of a PRA or a PSA comes from the simplification provided by event and fault trees, which in turn depends on the acceptance of the bimodal principle and of the assumption that effects combine or propagate in a linear fashion. Both assumptions are clearly visible in the event tree, the first as the binary branching through 'successes' and 'failures' and the second by the very nature of the tree representation, cf. Figure 7.2. (The same argument can be applied to all commonly used risk or safety assessment methods, including cause-consequence trees and more complex representations such as networks of various types.)

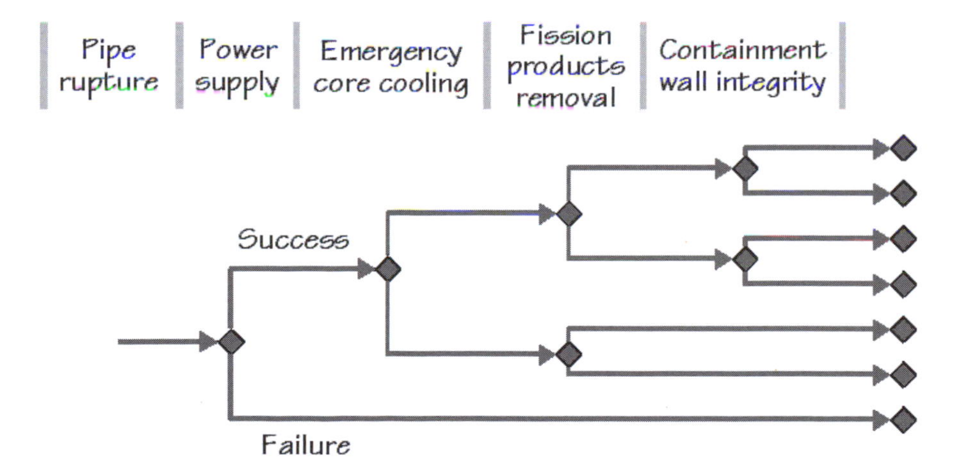

Figure 7.2: A PRA event tree

The event tree represents the possible ways in which the consequences of an initial failure ('pipe rupture') can propagate through a system, in this case a nuclear power plant. This is done by listing the critical steps or events necessary to avoid the adverse outcome, such as 'power supply,' 'emergency core cooling,' etc. For each step the probability of success or failure is assessed in order to calculate the combined probability of the different outcomes. If the failure probability of a step is not already known, it can be determined using, e.g., a subsidiary event tree, a fault tree, or expert judgement.

Without relying on this simplification it would be impossible to carry out a PRA/PSA for any large technological system within a reasonable budget and time. But the acceptance of these assumptions means that the thoroughness of the risk assessment is reduced. This trade-off, in combination with a lack of comprehensive data and the problems associated with modelling assumptions that cannot easily be quantified, is in practice compensated by increasing the safety margins, i.e., by using a more conservative criterion of acceptability. An interesting and probably unintended consequence of this is therefore that the efficiency of the system during operation in the end becomes limited by the lack of thoroughness of the risk assessment. (It is worth noting that this dilemma is not intrinsic to a PRA, but that it is the case for any kind of risk assessment.)

HRA as ETTO

Shortly after PRA was taken into use in 1975, something happened that forever changed the landscape of accident investigation and risk assessment in general, and of PRA in particular. This was the accident at the Three Mile Island nuclear power plant on 28 March 1979. The analysis of the accident made clear that it was necessary for risk assessments to go beyond a description of how the technological system functioned and include the effects of actions by the operators in the system, the human factor, as well. In terms of the commonly used tool of reliability engineering, i.e., the event tree, the question became how it would be possible to calculate the likelihood of a human action failure.

The solution was simple enough, namely to introduce human actions as nodes or steps in event and fault trees. The solution was understandable in terms of the urgent need to find a way to include the human factor in risk assessments. Yet in retrospect even this choice can be seen as an efficiency-thoroughness trade-off: it was better to begin sooner by using a method that seemed to work than to delay a response by thinking about how the problems should best be solved.

Human Reliability Assessment (HRA) was introduced as a complement to PRA to calculate the probability of a human action failure, or 'human error,' in places where it was needed. This quickly proliferated to 'human error' being used in practically every possible context – and in some impossible ones as well. As a result of that the notion of 'human error' has today become so firmly entrenched in the thinking about safety that it is taken for granted. The value of using it is therefore rarely – if ever – doubted.

To invoke the notion of 'human error' is unquestionably the most common ETTO in safety work. It is efficient because it provides a single and simple explanation; but it lacks thoroughness because the explanation is insufficient, if not directly wrong. That it is insufficient can be seen from the many proposals for detailed classifications or taxonomies of 'human error' – from the simple distinction between 'errors of omission' and 'errors of commission' to detailed lists of 'internal error modes' and 'psychological error mechanisms.' That it is wrong should hopefully be clear from the argumentation in this book. Humans always try to balance efficiency and thoroughness in what they do; indeed, they are expected to do so. It is only in hindsight, when the outcome is the wrong one, that the choice of efficiency over thoroughness conveniently is labelled 'human error.'

Clinical Trials of New Drugs

Going beyond the PRA/HRA issue, the importance of the trade-off between efficiency and thoroughness in risk assessment is illustrated in an exemplary manner by the way in which new drugs are approved for general use. The need of efficiency is due to both public health and commercial interests. The need of thoroughness is also obvious, since it is counter to the interests of everyone – be they patients, doctors, or pharmaceutical companies – to release a drug or medicine for common use if it either does not work as intended or if it has serious side-effects. This nevertheless happens every now and again, as demonstrated by the relatively recent case of Vioxx®, a non-steroidal anti-inflammatory drug that was used worldwide to treat patients with arthritis and other conditions causing chronic or acute pain. After the drug was approved as safe and efficient by the US Food and Drug Administration (FDA) in May 1999, it was prescribed to an estimated 80 million people worldwide. But on 30 September 2004, the drug was voluntarily withdrawn from the market because of concerns about increased risks of heart attack and stroke associated with long-term, high-dosage use. The FDA later estimated that Vioxx® had caused between 88,000 and 139,000 heart attacks (30 to 40 per cent of which were probably fatal) in the five years it was on the market.

The approval of a drug for general use is normally based on extensive clinical trials overseen by a national or international regulator. Such trials are guarded, amongst other things, by a code of conduct, the World Medical Association Declaration of Helsinki from 1964, which is a statement of ethical principles that provides guidance to physicians and other participants in medical research involving human subjects. One of the purposes of the code is to safeguard the health of the people who take part in such trials. The many safeguards implied by such codes or guidelines can obviously sometimes be seen as unnecessarily cumbersome and costly, hence be traded off for efficiency. In relation to the ETTO principle, one particular aspect is how long a clinical trial should take.

The purpose of a clinical trial is to determine whether a drug has the intended beneficial effects and no harmful side-effects. Since the drug – or a placebo – has to show effects on a disease, it is obviously important that it is allowed sufficient time to work, i.e., sufficient time for the intended effects to become manifest. For a test to be thorough, it should therefore take as long as it needs. In the case of a drug test, this is something that can be made more precise by using knowledge about the illness in question (e.g., whether somatic or psychological), to understand the way the drug works, etc. On the other hand, it is clearly desirable to have tests which do not last longer than necessary, for instance because the clinical trial itself costs money, because the delay in getting the approval means that potential revenues are lost, and because patients may eagerly await a new treatment. In order to regulate this issue, the guidelines of the US FDA state that clinical trials should last for around four to six weeks. Yet it is not unusual for drugs to take six weeks or longer to achieve maximum effectiveness or indeed to produce noticeable negative effects. Short trials may also fail to distinguish between transitory and permanent effects, and therefore fail to identify the people for whom the drug is initially helpful but later ineffective. The FDA further recommends that drug companies monitor trial participants for at least one week after they stop taking the drug, although it may take longer than that for withdrawal symptoms to emerge.

The issue here is not to make a value judgement about specific organisations and how they work, or to second-guess the motivation for setting acceptability criteria or for choosing one approval strategy or another. The issue is simply to point out how the ETTO principle can be found everywhere and how it can be used to understand why things are done in a certain way.

Quantification as ETTO

The denominator of risk and safety is the probability that something may go wrong. Common language often describes risks as, e.g., 'low,' 'moderate,' 'high,' or 'extreme,' and it is generally agreed that the risk represents a combination of the likelihood or probability of an outcome and the severity of the outcome. In qualitative terms, the outcomes may be described on a scale from 'certain,' 'likely,' 'possible,' 'unlikely,' and 'rare,' whereas the probability is expressed by a number, e.g., 2.8E-5.

Using the probability value is clearly an efficient way of talking about risks, not least because it provides a simple way of comparing different risks. Something that is 'possible' is clearly more risky than something that is 'rare,' but how much more? If instead we say that the risk of X happening is 4.1E-4 and the risk of Y happening is 2.8E-5, then we easily calculate that the risk of Y is 14.6 times larger than the risk of X. Such a comparison is however misleading, since the numerical value requires a detailed understanding of the context to make sense. Yet it is clearly much faster to compare two values than to compare the underlying investigation results and the detailed descriptions.

A good example of the problems with quantitative indicators is provided by the concept of Value at Risk (VaR) used in the world of finance. The VaR is calculated for a specific portfolio at a specific point in time, and represents the worst case loss that could be expected to be incurred from a given portfolio as a result of movements in identified risk parameters, over a nominated time period within a specified level of probability. It is clearly easy to compare two VaR values, but it is obviously also efficient rather than thorough. Indeed, the use of the VaR has been severely criticised by financial experts and scholars, not least in the analysis of the meltdown of the financial system in 2008.

Concluding Comments

The argument that risk assessment by necessity represents an efficiency-thoroughness trade-off applies not only to PRA/HRA but to other methods as well. Indeed, even new methods that try to overcome the limitations that come from the bimodal principle and the linearity assumption will have to trade off thoroughness for efficiency. It is quite simply impossible not to do so! The problems therefore lie less in the fact that a trade-off is made than in the way in which it is made. The dilemma, for risk assessment as well as for many other activities, is that the need to be efficient in the short run is in conflict with the need to be efficient in the long run, because the latter – almost counter-intuitively – creates a need to be thorough in the short run as well! If, for instance, a risk assessment or a test does not go into sufficient detail, the long-term efficiency of the process or product being assessed will suffer.

Given that any risk assessment must involve an efficiency-thoroughness trade-off, the question becomes how the ETTO principle can be used to provide a better understanding of risks. In other words, we need to know better how human and organisational performance is adjusted to the conditions, and to use that understanding to predict what may happen in the future. More concretely, we need to develop methods that make it possible to account for failures that are emergent rather than the result of cause–effect relations. One such method is the Functional Resonance Analysis Method (FRAM) already mentioned in the sources of Chapter 6. This method is based on four principles: (1) that successes and failures have the same underlying explanations, (2) that performance adjustments are ubiquitous and approximate, (3) that consequences are emergent rather than resultant, and (4) that the traditional cause–effect relation can be replaced by a resonance relation. The first three principles are consistent with the ETTO principle and have been mentioned several times in the previous chapters. The fourth principle explains the spreading of consequences through tight couplings as a dynamic phenomenon, rather than as a simple combination of causal links.

Sources for Chapter 7

Amalberti has described some very interesting ideas about the difference between ultra-performing systems and ultra-safe systems, where the latter are systems with risks lower than 10^{-6} per safety unit. A good introduction to ultra-safe systems is R. Amalberti, (2002), 'Revisiting safety and human factors paradigms to meet the safety challenges of ultra complex and safe systems,' in B. Wilpert and B. Fahlbruch (eds), *System Safety: Challenges and Pitfalls of Intervention* (Oxford: Pergamon). Another reference is R. Amalberti, (2006), 'Optimum system safety and optimum system resilience: Agonistic or antagonistic concepts?,' in E. Hollnagel, D. D. Woods and N. G. Leveson (eds), *Resilience Engineering: Concepts and Precepts* (Aldershot: Ashgate).

The idea that performance is affected by the conditions of work has been part of Human Reliability Assessment (HRA) from the very beginning. In the 1970s it was commonly referred to as Performance Shaping Factors. In the so-called first generation HRA methods, the performance conditions were assumed to influence the underlying 'human error probability.' In the second generation methods this view was revised, so that performance conditions were seen as the primary determinant of performance reliability.

The use of the terms sharp end and blunt end was probably introduced by James Reason's influential book on *Human Error* (Cambridge University Press, 1990) and later refined by D. D. Woods, L. J. Johannesen, R. I. Cook and N. B. Sarter, (1994), *Behind Human Error: Cognitive Systems, Computers and Hindsight* (Columbus, OH: CSERIAC). The basic idea is that most, if not all, of the failures made at the sharp end are determined by the working conditions and the nature of the tasks, hence by what other people have done at an earlier time and in a different place. Sharp end and blunt end are, however, relative rather than absolute terms, since clearly 'everybody's blunt end is someone else's sharp end.'

The 'small world' phenomenon is the idea that every person on the earth is connected to every other person through at most five other people. More generally, it is the idea that any part of even very large socio-technical systems is coupled to another part through a limited number of links or degrees of separation. The actual number of links is less important than the fact that this creates a way for consequences to spread in an unexpected, and usually uncontrolled, manner. According to Wikipedia, the idea is nearly a century old. The classical demonstration of the 'small world' phenomenon is by J. Travers and S. Milgram, (1969), 'An experimental study of the small world problem,' *Sociometry*, 32(4), 425–443.

The WASH-1400 report was produced by a committee of specialists under Professor Norman Rasmussen in 1975 for the USNRC, and is therefore often referred to as the Rasmussen Report. The WASH.1400 study established PRA as the standard approach in the safety-assessment of modern nuclear power plants. It does not seem to be available on-line.

The development of HRA has been treated by many books and papers. The change from first to second generation HRA methods was due to a severe criticism of the former by one of the leading practitioners at the time (E. M. Jr. Dougherty, (1990), 'Human reliability analysis – Where shouldst thou turn?,' *Reliability Engineering and System Safety*, 29(3), 283–299.) Comprehensive surveys of HRA methods and of the history of HRA can be found in, e.g., E. Hollnagel, (1998), *Cognitive Reliability and Error Analysis Method* (Oxford: Elsevier); or B. Kirwan, (1994), *A Guide to Practical Human Reliability Assessment* (London: Taylor & Francis).

Epilogue

Having argued for the omnipresence of the ETTO principle – and of ETTOing – the inescapable question is why whatever we do seems to require a trade-off between efficiency and thoroughness.

Performance adjustments are necessary whenever the situation or the working conditions are underspecified. The only exceptions to this are perfectly controlled situations where the possibility of disturbances and interruptions is negligible. One example is industrial work during the first decades of the 20th century, at least for believers in the principles of Scientific Management. Today more obvious examples are in the realm of rituals and drills, such as religious services or peacetime military performances. The changing of the guards, for instance, takes place with the precision and regularity of clockwork – and is indeed often described in those terms. But there are few other situations where people can perform as machines and still feel certain that the outcome will be correct. It is only when the controlling authority – the church, the military, etc. – provide conditions where there are no surprises and no time pressures that such performance is feasible. And from a safety point of view these cases are of minor interest, because a possible mismatch between performance and conditions will have limited consequences.

One consequence of underspecification is that it is uncertain both how long it will take to do something and how much time is available, because it is unknown if and when there will be an interruption of what is being done or a new demand to attention. It therefore makes sense to reduce the time spent on doing something, and through this create some slack or spare time. The slack is valuable both because it provides a buffer that can be used if an unanticipated demand arises, and because it can be used to think about performance, i.e., it provides an opportunity to improve and to learn. On a different level, being efficient rather than thorough may bring its own social and monetary rewards, at least until the lack of thoroughness catches up.

Single-Loop and Double-Loop Learning

Learning from failures involves the detection that a failure has occurred, the identification of the likely causes, and the formulation of appropriate counter-measures in one way or the other. (Note that even if the failure is correctly detected – which it normally is – an incorrect identification or analysis may lead to incorrect countermeasures, hence to an inefficient response.) We tend to think of this learning as something that happens directly, in the sense that the countermeasures are devised for the conditions that existed when the failure happened. In other words, the countermeasures permit the organisation to go on with its present policies or achieve its present objectives. This kind of error-and-correction process is called *single-loop* learning. It is, however, also possible to change the conditions themselves, for instance by improving the conditions for work. Since this is a change that may affect the more direct countermeasures, it is called *double-loop* learning. Using the words of Chris Argyris, double-loop learning 'occurs when error is detected and corrected in ways that involve the modification of an organization's underlying norms, policies and objectives.'

Single-loop learning is characteristic of situations where goals, values, frameworks and strategies are taken for granted. Any reflection is aimed at making the strategy (of error correction) more effective. Single-loop learning is thus fast, since it can focus on the concrete situation without the need of more elaborate analyses. Double-loop learning, in contrast, means that the role of the framing and learning systems, which underlie actual goals and strategies, is questioned. In the first case reflection is very concrete, in the second it is more fundamental and involves higher-level considerations and longer time spans. The feedback is therefore also both delayed in time and more generic in content.

Years of studies in social science show that the greater the need is and the more that is at stake, the more likely it is that individuals – and organisations – rely on single-loop learning. This is not really surprising since efficiency under most conditions is valued higher than thoroughness, but it does mean that less is learned from the failure than possibly could be. What is achieved is an immediate and possibly short-lived correction, rather than a more thorough assessment and a possible revision of the conditions for work. The difference between single-loop and double-loop learning is thus very similar to the ETTO principle.

From ETTO to TETO

To trade off between efficiency and thoroughness in order to get through the work-day is normal, necessary, and useful. However, as Resilience Engineering points out, it is not sufficient to be able to do something or to respond to the actual; it is also necessary to consider the critical, i.e., to monitor that which is or could become a threat in the near term. In other words, efficiency in the present presupposes thoroughness in the past, which paradoxically means that thoroughness in the present is necessary for efficiency in the future.

It follows from this argument that the ETTO principle requires a symmetric TETO or Thoroughness-Efficiency Trade-Off principle. Just as the ETTO principle can be interpreted as representing single-loop learning, the TETO principle can be interpreted as representing double-loop learning. The question, of course, remains, when one should put the emphasis on efficiency and when on thoroughness. For an organisation that question may not be too difficult to answer, since there are clear differences between the day-to-day operations and functions such as supervision, and learning. It is practically a definition of an organisation that these functions can be assigned to different parts or to different roles. For an individual it is more of a problem, since it is impossible literally to do two things at the same time. For an individual, the ETTO–TETO balance therefore becomes an issue of scheduling various activities, and of creating time enough for reflection. Individual intentions to maintain a balance, to be thorough as well as efficient, may nevertheless easily run foul of time pressures, information push, and information input overload and must therefore be supported by the organisational culture.

It is unrealistic in practice to require everyone to be thorough rather than efficient, even though this often seems to be a direct or indirect outcome of accident investigations when the recommendation is that people should 'follow procedures.' If everyone really tried to be thorough it would be devastating to productivity without necessarily improving safety, as illustrated by 'work-to-rule' situations. It is equally inadvisable to allow everyone to be efficient rather than thorough, since that definitely will put the system at risk, perhaps even in the short run. One solution is to insist on thoroughness – or TETO – in places where there is a transition between subsystems or sets of functions, and where this transition is known to be critical to both productivity and safety.

Such solutions must be based on a different approach to system analysis, one that focuses on couplings and dependencies rather than on failures and malfunctions.

The TETO principle should not be seen as the antagonist of the ETTO principle, but rather as a reminder that efficiency and thoroughness are relative rather than absolute terms. The trade-offs of thoroughness for efficiency that are made in short term may sometimes turn out to be counter-productive in the long term, in the sense that they reduce the efficiency of the system or production if seen over longer periods of time. System maintenance, be it of equipment or 'liveware,' may for instance be done less frequently than initially intended, due to production pressures, lack of resources, etc. This may gradually enlarge the distance between intended and actual system states, and in the worst case lead to a serious malfunction or even a collapse of the system, for instance as an accident. Conversely, a trade-off of efficiency for thoroughness in the short term may turn out to benefit efficiency in the long term. As a simple example, if tools always are returned to their original place after use, less time will be spent on looking for them later. This is so for the do-it-yourselfer as well as for a people working together in a work unit or a factory. (It is even the case in office environments!) TETOing can also be recognised in the principle of preventive maintenance, be it of humans, technologies, or organisations, in quality management, of safety culture, etc.

In order to survive in an underspecified world it is necessary to know why things happen. We need to know why things go right in order to ensure that we can produce the same outcome again. We should be careful not to succumb to the pressure just to continue doing what we do, rather than to pause and reflect. We also need to know why things go wrong, but should avoid falling into the trap of relying on routine investigations and stereotyped explanations. The argument throughout this book has been that things go right and go wrong for the very same reason, namely that people adjust their performance as if they made an efficiency-thoroughness trade-off. The formulation is deliberately careful, since it is not claimed that the human mind works in this way, or that we in any way understand the possible underlying cognitive or neural processes. More importantly, it is not even necessary to do so. It is sufficient that the ETTO principle has a practical or pragmatic value as a short-hand (or efficient?) description that improves our likelihood of muddling through in our self-created intractability.

Finally, the ETTO principle itself represents an efficiency-thoroughness trade-off. Using it as a universal explanation for human and organisational performance may be tempting, but it is wrong. If that happens, it should be a cause for concern.

Sources for Epilogue

Scientific Management, also know as Taylorism, was a theory of work that took the division of labour to its logical extreme. The core ideas of the theory were developed by Frederick Winslow Taylor in the 1880s and 1890s, and were fully described in his book *The Principles of Scientific Management* (Harper & Brothers, 1911). The first principles were to analyse tasks to determine the most efficient performance and then select people to achieve the best possible match between task requirements and capabilities.

The ideas about single- and double-loop learning were developed by Chris Argyris, currently Professor Emeritus at Harvard Business School, as part of his work on learning organisations. The quotation in the text is from C. Argyris and D. Schön, (1978), *Organizational Learning: A Theory of Action Perspective* (Reading, Mass: Addison-Wesley).

A clear example of the consequences of increasing maintenance intervals is described in R. Woltjer and E. Hollnagel (2007), 'The Alaska Airlines Flight 261 accident: A systemic analysis of functional resonance,' International Symposium on Aviation Psychology, 23–26 April, Dayton, OH.

Subject Index

ERAU - PRESCOTT LIBRARY